Analytical Chemistry: A Guided Inquiry Approach

Instrumental Analysis Collection

Developed through the ANAPOGIL Project

Project Directors

Juliette Lantz

&

Renée Cole

POGIL Project
Director: Richard Moog
Associate Director: Marcy Dubroff
Publication Liaison: Sarah Rathmell

ISBN: 9781-1-190-6509-8

Table of Contents

ANAPOGIL Project People

Leadership

Juliette Lantz, Project PI – Drew University, Madison, NJ
Renée Cole, Project co-PI – University of Iowa, Iowa City, IA

Consortium Members

Chris Bauer – University of New Hampshire, Durham, NH
Christine Dalton – Carson-Newman College, Jefferson City, TN
Anne Falke – Worcester State University, Worcester, MA
Shirley Fischer-Drowos – Widener University, Chester, PA
Caryl Fish – St. Vincent College, Latrobe, PA
David Langhus – Moravian College, Bethlehem, PA
Ruth Riter – Agnes Scott College, Decatur, GA
Carl Salter – Moravian College, Bethlehem, PA
Mary Walczak – St. Olaf College, Northfield, MN

Affiliated Authors

Kathleen Cornely – Providence College, Providence, RI
Dan King – Drexel University, Philadelphia, PA
Paul Jackson – St. Olaf College, Northfield, MN

Additional classroom testers and project participants

Carlos Garcia – University of Texas-San Antonio, TX
Gija Geme – University of Central Missouri, Warrensburg, MO
Gabrielle Haby – University of Texas-San Antonio, TX
Elizabeth Jensen – Aquinas College, Grand Rapids, MI
Delana Nivens – Armstrong Atlantic State University, Savannah, GA
David Paul – University of Arkansas-Fayetteville, AR
Josh Smith – Armstrong Atlantic State University, Savannah, GA

Additional Chemistry Education Research personnel on the project

Meagan O'Brien – University of Central Missouri, Warrensburg, MO (undergraduate researcher)
Jennifer Schmidt – University of Iowa, Iowa City, IA (graduate student)
Wendy Schatzberg – University of Iowa, Iowa City, IA (post-doctoral fellow)
Stephanie Ryan – Chicago, IL (consultant)

Advisory Board

Diane Bunce – Catholic University, Washington, D.C.
Jonathon Crowther – Ortho-Clinical Diagnostics, Inc., (Johnson & Johnson), NJ
Jeanne Pemberton – University of Arizona, Tucson, AZ

This material is based upon work supported by the National Science Foundation under Grant No. 0717492. Any opinions, findings, and conclusions or recommendations expressed in this material are those of the author(s) and do not necessarily reflect the views of the National Science Foundation.

Overview of project

The intent of the ANAPOGIL project is to provide a set of POGIL (Process Oriented Guided Inquiry Learning) materials that impart widely accepted analytical chemistry principles while engaging students as active learners, and facilitating their development in the process skills that are essential to analytical chemists in particular and valuable to scientists in general. These classroom activities can be used collectively in traditional analytical chemistry courses or used individually in a variety of other courses that include analytical chemistry concepts. To develop the initial materials, a team of ten analytical chemists from various institutions convened over a six-year period to write, test and assess these guided-inquiry materials with the full participation of two chemical education researchers as part of an NSF funded project (DUE # 0717492). During the latter years of the project, additional classroom testing was conducted at five additional institutions.

Acknowledgments

This collection of activities and supplementary materials is the result of the interactions and support of a large number of people and organizations.

- Thanks to the National Science Foundation (Grant # DUE-0717492) for its support of the project that has allowed us to accomplish so much more than we would have believed possible.
- Special thanks to the POGIL Project, and specifically Rick Moog, for getting us started on this path. This project was born out of interactions among consortium members at a POGIL National Meeting. The POGIL Project also provided the seed money for our first consortium meeting that helped us to write a successful grant proposal.
- Many thanks to the POGIL project staff who have helped us in transforming our activities into a professionally edited and presentable form that could be published.
- Thanks to our many colleagues in the POGIL community who have provided us with the opportunity to discuss our ideas with interested, stimulating, and dedicated colleagues. Many POGIL members provided guidance, insight, and support throughout the project.
- Thanks to the faculty who have used our activities in their classrooms, and special thanks to those who have provided us with feedback to continue to improve the quality of our activities.
- A great debt of thanks is due our students. They have provided their insights into the structure of the activities and helped us improve these materials.
- Drew University further supported the project through contributions of release time and computer equipment; these contributions are gratefully acknowledged.
- Finally, a special acknowledgment from JL and RC to the fabulous consortium members whose dedication and enthusiasm for the ANAPOGIL project never wavered over the several years of continued consortium tasks. This is was an incredibly hard working, generous, creative, organized, and downright fun group of analytical chemists; their collaborative efforts were key to the success of this project.

Overview of materials

The ANAPOGIL consortium has written activities that fall into six main categories typically covered in an undergraduate analytical chemistry curriculum: Analytical Tools, Statistics, Equilibrium, Electrochemistry, Spectrometry, and Chromatography and Separations. The activities were assembled into two collections. The Quantitative Analysis Collection contains 28 activities that span the range of topics likely to be covered in a semester-long quantitative analysis course. This includes the entire sets of Statistics, Equilibrium, and Electrochemistry activities and the foundational activities in the Analytical Tools, Spectrometry, and Chromatography categories. The Instrumental Analysis Collection (20 activities) contains the full sets of Electrochemistry, Spectrometry, and Chromatography, as well as the activities for the more advanced topics in the Analytical Tools category. Two of the statistics activities that pertain most strongly to Instrumental Analysis were also included. The Instrumental Collection does not yet include a full set of instrument-specific activities that might commonly appear in an Instrumental analysis course; for instance, coverage of FTIR, NMR, MS and surface techniques is not yet available. We hope to add these topics in future editions.

There is significant overlap of the Quantitative and Instrumental Collections (14 activities appear in both books) reflecting the amorphous boundaries of these courses at different institutions. Wiley Custom Select provides custom publishing options for all the activities in the entire ANA-POGIL collection to allow instructors to optimize instructional materials to fit particular course needs.

All published activities have been reviewed and classroom tested by multiple instructors at a wide variety of institutions. Each written activity includes content and process goals as well as a list of prerequisites. This information serves to help instructors incorporate the activities into their courses. The topical treatment of concepts, as well as advanced coverage of analytical principles, can be found in the application section of each activity. Instructors can tailor coverage for a particular course by assigning different combinations of application questions.

All written materials are modular in nature, and they include cues for team collaboration and self-assessment. All materials follow the constructivist learning cycle paradigm. In the exploration phase of each activity, a team of students is presented with a model, which may be a figure, a data table, a graph, an experimental procedure, a problem-solving methodology, a schematic of a device, or any combination of these. Directed questions encourage students to deeply explore the model, especially the fundamental relationships or concepts embedded in it. Concept development is supported by convergent questions, which guide students to make connections and form hypotheses. These questions model the questions scientists ask in attempting to understand new information. Term introduction typically occurs at this phase, to help students identify their newly developed concept. Students are often asked to clearly articulate the concept using complete sentences. In addition to developing communication skills, clear expression of the concept is essential for true understanding of the concept. In the concept application phase of the activity (which completes the learning cycle), students are asked to apply the concept to new situations. Skill exercises are one or two step exercises that provide repetition of the key questions in an identical or similar context as the model; they help to build confidence and strengthen understanding. Finally, students are asked to apply their new knowledge to solving more advanced problems. These problems require higher level thinking skills such as synthesis or analysis, may involve a multi-step approach integrating several concepts or skills, may require assumptions, and are generally more real world in nature.

In short, these activities will take students far down the path to becoming analytical chemists. Going beyond content mastery, students should gain a strong sense of what an analytical chemist does, and more importantly get to experience many aspects of how an analytical chemist does it. These activities directly scaffold every laboratory skill and technique an analytical chemist should master; moreover, the critical thinking skills and information processing can greatly empower students to choose effective data analysis routes and make sound, defensible experimental decisions.

Overview of classroom usage

These activities are mainly designed to be used in the classroom and have been implemented in settings with class periods ranging from 50 to 75 minutes. The intent is for students to work through the activities in structured Learning Teams of 3-4 students with the instructor actively facilitating student engagement and learning during class. The activities have been listed in the Table of Contents in an order that many instructors have found useful, but each activity lists the assumed prerequisite knowledge to allow the most flexibility on the part of the instructor for sequencing the use of the activities. The materials do not replace a textbook, rather they are designed to enhance the learning experience in the classroom component of the course. However, due to the learning cycle/guided inquiry design of the activities, it is recommended that students complete the activity and subsequently read the corresponding sections of the textbook. Instructors may also assign problems from the text as part of homework assignments.

Each instructor naturally implements activities in ways that are unique to that individual and institution, but we have found some common approaches that lead to more successful outcomes. We have found an implementation that includes mini-lectures and whole-class discussions in addition to the small group work to be very effective. Students may initially move slowly in this learning paradigm, but become more adept as activity usage builds and they gain insights into learning strategies and confidence in themselves and teammates. For this reason, it is important to use activities on a regular basis (daily, weekly, biweekly, etc.) so that students become accustomed to this learning environment and their role in it. We have created an instructor's guide with detailed tips on implementing the activities (including pacing guides.) Suggested implementation strategies include assigning initial directed questions in the activity as pre-class work to be discussed at the beginning of class, using the application questions as homework, providing mini-lectures to introduce background material or to address common sticking points, using whole-class discussions to compare student responses and elaborate on ideas, and providing a summary at the end of the class to highlight the learning objectives addressed. We also highly recommend that potential users attend a POGIL workshop to gain a deeper understanding of the philosophy and facilitation of guided inquiry activities.

It is important that students get some feedback on content mastery gained through use of the activities– either through classroom summaries, graded questions, or facilitator review of student work as it unfolds in the classroom. Instructors do not typically grade the activities themselves, but often focus on particular application problems and perhaps a small subset of key questions, particularly those that are summative in nature. These can be graded individually or for the group, using a Learning Team report. It is also crucial to explicitly address process skill mastery–which entails some form of evaluating student team efforts, communication skills, information processing, self-assessment (metacognition), and critical thinking as well as the more traditional problem solving strategies. The instructor's guide includes examples of several classroom strategies to assess process skills, including assessment reports and exam questions that address process skills as well as content mastery.

A two-part instructor's guide is available for these activities. Part A focuses on answer keys and instructor tips that are specific for each activity, including common sticking points and suggestions for natural break points to fit timing needs. Part B is a more general guide, with a larger focus on overall implementation and assessment strategies, including classroom procedures and materials that have been used successfully. It includes chapters on: (1) Planning for implementation and using Learning Teams, including managing expectations and structuring the course; (2) General implementation strategies, particularly for implementation in an upper level course; (3) Facilitation strategies and materials such as buy-in activities, recorder's sheets, and strategist reports; (4) Course assessment strategies, particularly for process skills; and (5) Perspectives from experienced users including sample syllabi, course descriptions, and frequently asked questions and frequently felt feelings experienced by instructors during their first implementation (and beyond).

Sample Preparation

Learning Objectives

Students should be able to:

Content

- Evaluate sample preparation methods for completeness.
- Determine the concentration of analyte in the sample from the concentration in the digestate.
- Explain how to assure sample integrity with respect to contamination and accuracy.

Process

- Form an understanding of the accuracy of a measurement. (Problem Solving, Critical thinking)
- Analyze methods for robustness and sources of error, and overall quality. (Critical thinking)

Prior knowledge

- Facility with solution dilution calculations.
- Ability to calculate standard deviation.

Further Reading

- D.C. Harris, *Quantitative Chemical Analysis,* 7th Edition, 2007 W.H. Freeman: USA, Chapter 28, pp. 644-659.

Author

Anne Falke

Consider this...

Copper mining has been an important industry in the United States for the past two centuries and remains so today. The Anaconda Copper Mine, located in Montana, provided much of the country's copper from 1892-1973 (EPA, 2011). Until 1915, the waste from the mining process, which contains substantial amounts of copper, was dumped into surrounding rivers. Copper is listed by EPA as a "priority pollutant" (Irwin, et al., 1997). It tends to bind strongly to sediments in the form of copper sulfide. Normal concentrations in soil range from 5-70 mg/kg (ELC, 2008). In 2003 the EPA found elevated concentrations of copper in the Missouri River sediments around the Anaconda mine (EPA, 2011). The site was declared a Superfund site in March of 2011.

In a hypothetical analysis of the sediments downstream from the mine, three different analytical labs were contracted to determine the copper content in sediments. A sediment sample was collected, mixed, divided into three portions, one of which was sent to each of the analytical labs.

The analyst from Lab A took three samples of about 1g each from the batch of sediment, placed them in three different vials, added 10 mL HCl to each vial, and heated them for 2 hours. The resulting solutions were decanted from each vial, transferred to three 25 mL volumetric flasks and diluted to the mark with distilled water.

The analyst from Lab B took the batch of sediment, ground it into fine particles with a mortar and pestle, and then placed three samples of about 1g each into Teflon bombs (a thick-walled Teflon container with a tight-fitting cover that can withstand high pressure) for digestion. The analyst then added 10 mL of aqua regia to each sample. Aqua regia is a mixture of HCl and HNO_3. The two acids together are more corrosive than either alone. The analyst sealed the bombs and microwaved them for 10 minutes. Microwaving samples in Teflon bombs allows for high pressure, high temperature digestion in a short amount of time. The resulting solutions were transferred to three 25 ml volumetric flasks and diluted to the mark with distilled water.

The analyst from Lab C also prepared three samples using the aqua regia bomb digestion following the same process as the analyst from Lab B.

All analysts use graphite furnace atomic absorption spectrometry (GF-AAS) to analyze the samples. GF-AAS is a technique especially well-suited for trace analysis of metals in aqueous samples.

Key Questions

1. Diagram the process for Labs A and C described above following the pattern established be ow for Lab B.

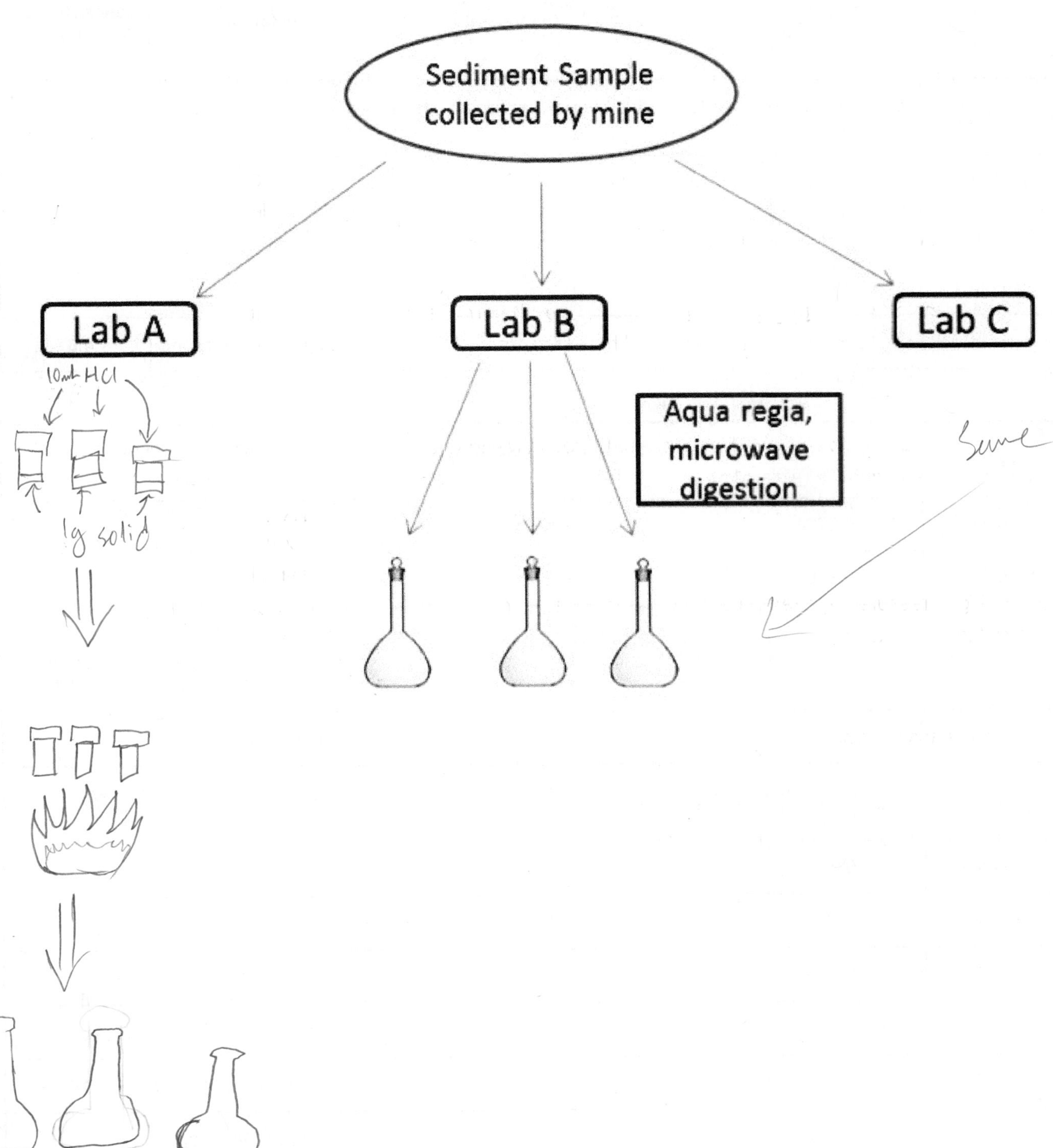

Consider this…

Table 1 *Copper concentrations in the sediment as determined by each of the analysts.*

	Sediment Sample Mass (g)	Copper concentration in digestate[a] (μg/mL)	Mass copper in digestate (μg)	Copper concentration in soil (μg/g)
Lab A	Digested samples are clear with a layer of black and white particles on the bottom of the 25-mL volumetric flask.			
Sample 1	0.9838	0.134	3.35	3.41
Sample 2	1.0049	0.245	6.13	6.10
Sample 3	1.0983	0.198	4.95	4.51
			Average concentration	**4.67**
			Standard Deviation	1.35
Lab B	Digested samples are clear, with only a few colorless to white particles on the bottom of the 25-mL volumetric flask.			
Sample 1	0.9794	0.765	19.1	19.53
Sample 2	0.9486	0.759	19.0	20.00
Sample 3	1.0273	0.750	18.8	18.25
			Average concentration	**19.27**
			Standard Deviation	0.906
Lab C	Digested samples are clear with only a few colorless to white particles on the bottom of the 25-mL volumetric flask.			
Sample 1	0.9845	0.726	18.2	18.44
Sample 2	0.9623	0.733	18.33	19.05
Sample 3	0.9816	0.728	18.2	18.54
			Average concentration	**18.67**
			Standard Deviation	0.327

[a]Digestate is defined as the solution resulting after preparing the sample for analysis.

Key Questions

2. What is the final volume of each of the samples prior to GF-AAS analysis?

25mL

3. Notice that the mass of copper in the digestate for Lab A, Sample 1 has been calculated for you. Determine the equation used to calculate the mass of copper in the digestate and write it here.

mass Cu = [Cu] in digestate x Volume of Sample

4. Have one member of the group calculate the mass (μg) of copper in each 25 mL digestate solution for Lab A, another member calculate the mass of copper in each 25 mL digestate solution for Lab B, and a third calculate the mass for Lab C. Write these numbers in the appropriate spaces in Table 1. Compare the results for each lab. What do you notice?

Lab A proposes a much lower concentration using much less precise data compared to the other two labs.

5. Notice that the concentration of copper in the sediment has been calculated for Lab B, Sample 2. Determine the equation used to calculate the concentration of copper in the sediment from the mass of copper in the digestate and write it here.

$$[Cu] \text{ in soil} = \frac{\text{mass Cu in digestate}}{\text{sediment sample mass}}$$

6. Have each member of your group take one of the labs, calculate the copper concentration in the soil for each sample and write it in the appropriate location in Table 1.

7. Calculate the averages for each analyst. Do they match the given averages?

Yes,

8. Calculate the standard deviations and write them in the appropriate box in Table 1.

9. Write a general equation to calculate the concentration of copper in each sample from the concentration in the digestate.

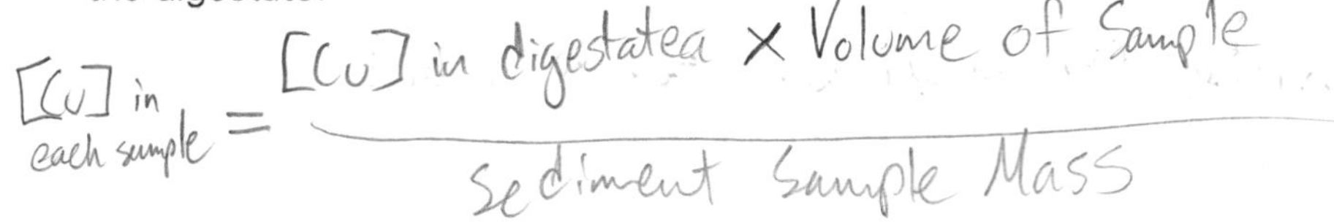

10. The three labs used two different sample preparation methods. List at least two differences between the methods.

1) Acid solution used for digestion

2) Heating time / Method.

3) Lab A didn't grind the sample.

11. Individually, examine the copper concentrations and standard deviations each lab found in their samples. How do the sets differ with respect to a) the analyte concentration and b) the precision?

Lab A proposes a much lower concentration and uses much less precise data than the other two labs, which had closer conclusions for [Cu] in the soil.

12. Do any of the differences noted in your answer to Q10 provide evidence for the results you see in your answer to Q11? Explain.

The aqua regia mixture appears to be important for digestion of the sediment sample. Lab A did not use this mixture, which may account for this discrepancy.

13. Based on your answer to Q12 summarize the important considerations when preparing samples for analysis?

- Strength of acid for digestion. Must be very corrosive.
- Grind sample for maximum surface area

14. In your group, brainstorm some experiments you could use to determine if a sample is completely digested. Think about how you can ascertain the portion of copper retrieved from the sample. As a group, propose a scheme for testing the completeness of a digestion.

- Use different concentrations of acids of different strengths until one reaches a
- Purchase a standard sample with a known concentration
  ↳ test for concentration and evaluate method.

Consider this...

All labs tested the accuracy of the method by analyzing a Standard Reference Material (SRM). SRMs are real samples that contain concentrations of many analytes which are so well characterized that they are certified by the National Institute of Standards and Testing. SRMs are available in almost any imaginable sample matrix (sea water, river water, estuarine water, bituminous and anthracite coal, various types of sediment, etc.). The sediment SRM used by the analysts is certified at 4.63 ± 0.04 μg/g copper in the solid sample. All three labs followed the same procedure for the SRM analysis as they did for their samples. Analysis for copper by GF-AAS yields the following results:

Table 2 *Concentration of copper (μg/g) in the sediment SRM as determined by the analysts.*

	Lab A	Lab B	Lab C
Trial 1	1.36	4.95	4.59
Trial 2	1.42	4.88	4.64
Trial 3	1.31	5.02	4.61
Mean	1.36	4.95	4.61
Standard Dev.	0.06	0.07	0.03
% Recovery	29%	107%	100%

15. What concentration of copper should the analysts detect?

4.63 ± 0.04 mg/g

16. Calculate the mean and standard deviation of the analysts' results and place the numbers in the appropriate spaces on the table.

Percent Recovery is the amount of the analyte detected compared to the theoretical amount present: concentration of analyte detected / theoretical analyte concentration x 100%

17. Calculate the percent recovery for each lab and place the numbers in the appropriate spaces on the table.

18. As a group, discuss the results in Table 2. Summarize your observations.

Labs B and C performed a more complete digestion of the sediment sample

19. Based on the results presented in Table 2, what can you say about the digestion process used by Lab A?

Incomplete

20. Someone suggests that the glassware in Lab B must be contaminated. Support or refute this statement with evidence from the data above.

Contamination is possible, given that the % recovery exceeded 100%

Consider this...

Fortunately the analysts thought ahead and prepared several different blanks:

Method blank – All reagents were added to the digestion bomb with no sediment and analysis was carried out identical to samples.

Reagent blank – 10 mL of aqua regia was added to a 25-mL volumetric flask and diluted to the mark with deionized water and analyzed as a sample.

Instrument blank – Plain deionized water is analyzed as a sample.

Table 3 ***Results of blank analyses.***

		Copper concentration (μg/mL)		
	Components	**Lab A**	**Lab B**	**Lab C**
Instrument Blank	DI H_2O	0.001	0.002	0.002
Reagent Blank	10 mL aqua regia, 15 mL H_2O	0.001	0.036	0.001
Method Blank	All reagents	0.003	0.034	0.002

21. Using the information provided, identify the components in each blank. Write your answers in the spaces provided in Table 3.

22. As a group, discuss the results. Record your thoughts. Does Lab A appear to have a contamination problem? Lab B? Lab C?

No, for A. Lab B screwed up. Lab C is cool.

23. What is the source of the contamination? As a group, discuss the source of the contamination and propose a solution to the problem.

The sample of aqua regia used by Lab B must be contaminated.

24. If the copper concentration for Lab B for the reagent blank had been very close to zero, what do you think would be the source of the contamination? Explain your reasoning.

The teflon bomb would be contaminated.

25. Look back at your answer to Question 18. With the additional evidence provided in Table 3 do you still agree with what you wrote? If not, how would you modify your answer?

Perhaps Lab B didn't have a complete digestion of the sediment, and was simply contaminated.

26. As a group, list three things you could do to assure minimal contamination in the sample prep and analysis.

Wash glassware, use clean and unused glassware, clean your instruments

27. You are the mine company manager. Which lab will you use for future analyses and why?

Lab C. Lab A didn't perform a complete digestion, and Lab B used contaminated reagents

Analysts need to assure the purity of all reagents, including distilled water, and the cleanliness of all equipment used. This is especially important when performing trace analysis. In addition to the blanks prepared in the lab, field blanks must be collected to assure that contamination is not picked up in the sampling process.

28. Brainstorm: How do you think a field blank is collected and treated? How is this different from a method blank?

Distilled water is run through an outside field and collects components for a sample. Method blanks include reagents used for treatment of the sample.

29. List two ways this activity has increased your understanding of sample preparation.

1) Many methods exist to test analytes, but some are more effective, and others may require improvement
2) Getting "accurate" results that one hopes for does not mean they truly are so. Compare with blanks.

30. List two tests you can use to determine whether or not a data set is reliable.

1) standard deviations
2) Percent Recovery using a standard.

Application

Homework

31. You have been hired by an analytical firm to set up the sample processing procedure for mercury in several large (greater than 10 kg) fish. List and explain several things you must consider when carrying out the digestion process.

32. Often it is desirable to extract the analyte from the matrix before analyzing. Extraction removes many compounds that may interfere with the analysis. One extraction technique useful for analysis of trace concentrations of metals in water is co-precipitation with some sort of chelating agent like Ammoniumpyrolidine-dithiocarbamate (APDC). The APDC forms an insoluble complex in water with metals. The precipitate is removed from the solution by filtration. The complex can then be dissolved in concentrated nitric acid with heat and diluted to a known volume with distilled water. Copper is extracted from 200.0 mL of riverine water by complexation with APDC, and the resulting precipitate is dissolved and diluted to 5.00 ml. The digestate is analyzed and found to contain 4.39 μg/ml Cu. What is the copper concentration in the river water? Locate information regarding ambient copper concentrations in natural systems and contamination levels. What can you say about the concentration of copper in this sample?

33. You have to prepare leaf samples for analysis. It is important that you dissolve as much of the leaf material as possible to extract the analyte. Propose a digestion process. Explain why you chose this method. (You may want to use your text to help you answer this question.)

34. SRMs are materials such as water, soil, plastics, blood, tissues, which are certified by the National Institute of Standards and Technology to contain specific concentrations of many analytes. Given that SRMs are real samples, how are the analyte concentrations known with such assurance?

Tinto, Rio; Kennecott Utah Copper website, http://www.kennecott.com/ 2011, accessed 5/20/11.

Environmental Literacy Council (ELC), Copper (2008), http://www.enviroliteracy.org/article.php/1029.html, accessed 5/20/11

Irwin, R.; van Mouwerik, M.; Stevens, L.;Seese, M.D.; and Basham, W., Environmental Contaminants Encyclopedia, Copper Entry, National Park Service, 1997, http://www.nature.nps.gov/hazardssafety/toxic/copper.pdf, accessed 5/20/11.

http://yosemite.epa.gov/opa/admpress.nsf/0/99138DEB5939D66D8525784D0076C9E0, accessed 6/2/2011.

Instrumental Calibration

Learning Objectives

Students should be able to:

Content

- Use a correlation coefficient to ascertain the fit of data to a calibration curve
- Determine the minimum detectable concentration and lower limit of quantitation and distinguish between these.
- Determine the linear dynamic range from a set of calibration data.

Process

- Interpret tabulated data. (Information processing)
- Produce an xy graph of data. (Graphing)
- Interpret calibration curves. (Information processing)

Prior knowledge

- How to produce and interpret calibration curves.
- The meanings of linear regression, correlation coefficient, accuracy and precision.

Further Reading

- Harris, Daniel C. 2007. "Quality Assurance and Calibration Methods." In Quantitative Chemical Analysis, 7th ed., pp.78-87. New York: W.H. Freeman.
- Skoog, Douglas A., F. James Holler, and Stanley R. Crouch. 2007. "Introduction." Principles of Instrumental Analysis, 6th ed., pp.11-22. Belmont, CA: Thomson Brooks/Cole.

Authors

Anne Falke and Shirley Fischer-Drowos

Consider this...

In the real world, you really don't know the concentration of analyte in your samples. There are tools analytical chemists use to ascertain the precision and accuracy of their methods. Generally, you need to understand the chemistry of the analyte and the matrix.

Consider the following analysis of silver in drinking water samples. The researcher starts out determining the limitations of the method by analyzing a wide range of calibration standards. The results are below:

Table 1

Ag (ppb)	Signal
0.00	0.000
2.00	0.013
10.0	0.076
20.0	0.162
30.0	0.245
50.0	0.405
75.0	0.598
100.0	0.748
125.0	0.817
150.0	0.872

Key Questions

1. Looking at the data, what do you notice about the relationship between the Ag concentration and the instrumental signal?

   The signal strength increases with silver concentration

2. Have one member of the group plot the signal (y-axis) vs. concentration (x-axis) on an Excel spreadsheet. Describe the pattern. *[Graphs may be supplied by your instructor. If graphs are supplied, you do not have to plot the data just look at Figure 1]*

3. Using Excel, insert a "best fit" line, often called a trendline in spreadsheet programs. Highlight the linear trendline. While you have that window open, check "Display R-squared value on chart" (R^2 or correlation coefficient is a measure of how well the data points fit the trendline. The closer R^2 is to 1 the better the fit). What can you say about the fit of the data to the trendline? *[Again, this question refers to Figure 1 on supplied graphs]*

The R^2 value is somewhat close to 1, but ideally it would be no less than $R^2 = 0.99$. This trendline doesn't fit the data very well.

4. As a group discuss how well the data fits the trendline. Do you notice some pattern to the points that do not fit? Describe the pattern.

The calibration curve does start to level off, starting at 100ppb. It starts linear, but the trend starts to curve and slope downward.

In many types of instrumental measurements, instrumental signal is no longer linear at high analyte concentrations. This may be due to chemical effects or to the limitations of the instrument.

5. What leads you to think that the data presented above is not linear at high concentrations?

The curve starts to slope downward starting at 100ppb. At any higher concentrations, the points appear to continue to slope downward.

At this point, the analyst would examine the data and determine at what point the data begins to deviate from linearity. Divide the work among group members. Have one member plot concentrations from 0 – 125.0 ppb, the second plot 0 – 100.0 ppb, the third from 0 – 75.0 ppb and so on until there are six different graphs for this data. Be sure to include trendlines and correlation coefficients. Format the data labels so that you have at least 5 decimal places in the correlation coefficient and trendline equation. Regroup and compare your graphs. [If graphs are supplied refer to Figures 2-6].

6. Looking at all of the graphs, what does the correlation coefficient tell you about the linearity of the data?

This coefficient describes how closely the data points fit the trendline of "best fit."

7. What happens to the correlation coefficient as data points at higher concentrations are removed?

The coefficient starts to improve and near closer to 1.

It is preferable to work in the region of the curve where the analyte signal is directly proportional to the analyte concentration. Analysts generally set a minimum acceptable correlation coefficient for a calibration. This varies with instrument and regulatory agency. Typical minimum acceptable correlation coefficients range from 0.995 to 0.9999.

8. As a group, decide which plot provides the best fit of the data without severely limiting concentration range over which the calibration can be used. Explain your answer.

The plot of 0-100 ppb provides the best fit of data. It provides an extensive range of data and the R^2 doesn't fall below 0.995.

9. What is the correlation coefficient for the selected graph?

R^2 = ~~0.9975~~ 0.99746

10. Look at the graph with the correlation coefficient just over 0.995. What do you notice about the last datapoint? Does this seem to be a result of random error or is it the start of rollover (negative deviation from linearity)?

The last data point does start to slope under the trendline. I think it's the start of the rollover. Some data points sit above the trendline, but this one falls below, which signals the start of the rollover.

11. a. Have a member of your group calculate the concentration of a sample with an analytical signal of 0.532 on the curve covering 0 – 100 ppb. Have a second member calculate the concentration of a sample with an analytical signal of 0.532 on the curve covering 0 – 75 ppb. As a group, evaluate how these numbers compare. What is the percentage error introduced using the curve with a hint of rollover?

0 - 100 ppb : 68.3 ppb w/ 0.532

0 - 75 ppb: 66.6 ppb w/ 0.532

% error 3% compared to 0-75 ppb graph

b. Discuss the error with your group. Do you think it is acceptable? Why or why not?

I think this error is acceptable. Doesn't seem too high.

This could be any combination of systematic or random error, but it doesn't seem enough to blame the scientist or the instrument

12. Based on the answers to the previous questions, decide as a group which is the best curve to use for this analysis. Explain your answer.

We're choosing 0-100. It provides a greater span of data and only begins to level off.

13. Based on the curve your group decided upon, what is the highest concentration standard that is still has a reasonably linear signal?

100 pb seems about the highest limit for analysis.

The value determined above is the upper limit of analysis. Any sample with a concentration greater than the upper limit cannot be reliably analyzed using the established calibration curve. For this reason it is desirable to choose the upper limit so that the curve is acceptably linear, but the available range is not excessively limited.

Consider this...

To continue establishing instrumental limitations the analyst carries out the following analysis by graphite furnace atomic absorption spectrophotometry. Note that the calibration is identical to the one established in the first part of this activity.

Table 2

Ag (ppb)	Signal
Calibration Standards	
0.00	0.000
2.00	0.013
10.0	0.076
20.0	0.162
30.0	0.245
50.0	0.405
75.0	0.598
100.0	0.748
Detection Limit Analysis	
0.00	0.003
0.00	0.007
0.00	0.004
0.00	0.006
0.00	0.008
0.00	0.004
0.00	0.005
2.00	0.012
2.00	0.017
2.00	0.015
2.00	0.013
2.00	0.019
2.00	0.017
2.00	0.016

Notice the information that follows the calibration curve data. The analyst has analyzed 7 replicate blank samples (no silver in the sample) and 7 replicate samples with a low concentration (1-5 times the detection limit) of silver. This is done to determine the minimum concentration of analyte that may be detected (analyte is present, but not necessarily at a concentration great enough to provide data reliable enough to quantify) and the minimum concentration that is quantifiable (present in a high enough concentration to measure with reasonable accuracy).

14. Discuss the difference between detection and quantitation with your group members. Write a sentence or two identifying that difference.

Detection is simply an identification of a signal present. Quantitation is the instrument's ability to accurately and precisely measure a concentration by measuring a signal.

15. Which should be the larger number? Why?

Quantitation should be higher, especially w/ larger concentrations, because it is more than simply accounting for the fact that a signal exists.

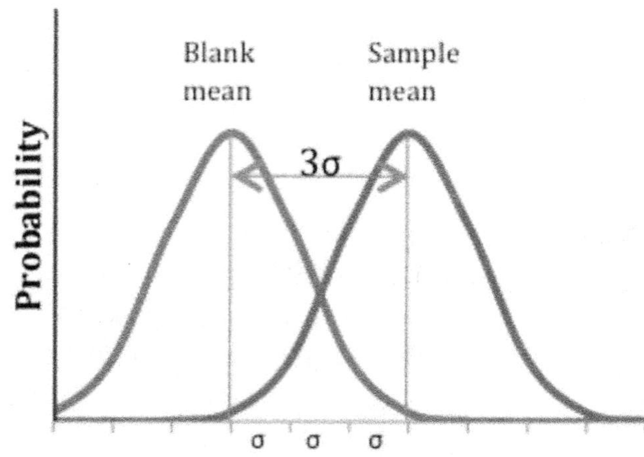

The variation in replicate analysis is a function of the randomness of the instrumental signal (noise). The analyst must be able to separate a real signal from the noise. Therefore, increased noise increases the limits of detection and quantitation.

The figure above represents the distribution of blank data and sample data. Note that the sample mean is 3 standard deviations 3σ greater than the blank mean. About 1% of the distribution of the blank data is greater than the sample mean. The minimum detectable signal is defined as the average signal for the blank + 3 times the standard deviation (s) of the low level sample signals. Half of the distribution for the sample data lies below the sample mean. What this means is that there is about a 1% chance that a blank will be detected as a real sample and a 50% chance that a sample with a mean at the detection limit will be viewed as a blank.

16. Calculate the average signal for the replicate blank analysis.

0.005

17. Calculate the standard deviation for the replicate 2.00 ppb standards.

0.00244

18. Calculate the minimum detectable signal for the data above.

0.005 + 3(0.00244)

= 0.0123

19. Using the calibration curve chosen in question 12, individually calculate the concentration that corresponds to the minimum detectable signal. Compare your answer with your group members. When you have a single value that you all agree on write it here.

The concentration that corresponds to the minimum detectable signal is defined as the lower limit of detection (LLOD) (also called minimum detectable concentration or detection limit). *If the intercept for the calibration curve and the average blank signal are both approximately zero* as is typically the case in atomic absorption spectrophotometry, the LLOD can be estimated from 3 times the standard deviation of the low concentration signals divided by the slope of the calibration curve.

LLOD = 3s/m

20. Calculate the LLOD using 3s/m method to prove that the two calculations yield approximately the same concentration (of the same order of magnitude).

$$LLOD = \frac{3(0.00244)}{0.0077}$$

21. What are the units for the LLOD?

ppb

22. As a group, summarize the calculations for detection limits:

a. Minimum Detectable Signal

0.0123, lowest possible signal detectable from the lowest concentrations

b. Lower Limit of Detection (exact method)

0.781 ppb, lowest possible concentration detectable associated with the calculated minimum detectable signal.

c. Lower Limit of Detection (estimated method)

0.951 ppb, lowest possible concentration detectable

23. In what situations would you use each of the above?

In cases where only detection is necessary (ex. is cocaine present?) only MDS is necessary.

The lower limit of quantitation (LLOQ) is defined as the concentration which corresponds to the signal that is 10 times greater than the noise.

24. Write an equation to determine the LLOQ.

~~LLOQ = 0.005 x10~~

$LLOQ = X_{blank} + 10(standard)$

25. What is the signal for the LLOQ for the data set above?

$LLOQ = 0.005 + 10(0.00244)$

$LLOQ = 0.0294$

26. Using the calibration curve established in Q12 determine the concentration corresponding to the LLOQ.

~~3 ppb~~ 3.037 ppm

27. Identify the assumptions used in the estimation LLOQ calculation.

The noise remains relatively constant or we're using the same instrument.

As with the LLOD, for methods that produce a calibration with an intercept of approximately zero, the LLOQ can be estimated:

$$LLOQ = 10s/m$$

slope of calibration curve

28. As a group, compile a clear set of directions for your lab technician describing how to determine the LLOD and LLOQ for a new instrument.

- Take multiple signal measurements of multiple blank samples
- Take multiple signal measurements of multiple low level concentrated samples.
- Take signal data of samples with varying concentrations and determining the calibration curve's equation of best fit
- Calculate an average signal for the blank analysis.
- Calculate a standard deviation for the low concentration analysis
- Calculate the minimum detectable signal for the instrument as the average blank signal plus 3X the standard deviation of the low concentration analysis. Then calculate the concentration of that signal based on the original calibration curve (LLOD)
- Calculate the LLOQ as the average blank signal + 10X the st. dev. of the low concentration analysis

29. What is the linear range (upper and lower limit) for the set of data in Table 2?

3.037 ppm ⟶ 100 ppm

The linear range is defined as the range over which the instrumental signal is proportional to the concentration of the analyte. The lower end of the linear range is the LLOQ and the upper limit is the highest concentration standard for which the signal is still linear. The linear range is expressed in concentration terms.

30. Since the linear range encompasses the analyte concentrations for which the signal is linear, a signal out of that range cannot be reliably quantified. What can you do to reliably analyze a sample that yields a signal of 0.936 on this curve?

31. What can you do to get reasonable quantitation of a sample that initially yields a response of 0.010 with this curve?

32. List and explain two ways that your understanding of the principles and use of calibration curves has improved.

33. How does graphing tabulated data enhance your ability to interpret it?

34. Given the following set of emission data, determine linear range of quantitation for the analysis. Calculate the LLOD using both the exact and estimated methods and compare to the lower limit of your linear quantitation range.

Application

35. The following samples were analyzed for silver. To each digestion flask 200 mL aliquots of 10 different drinking water samples were added. Silver was extracted by a co-precipitation procedure and collected by filtration. The precipitate was digested in 200 µl hot concentrated nitric acid and diluted to 5 mL in a dilute phosphate solution. Standards were prepared according to the following table. The samples were analyzed by graphite furnace AAS

[Pb] (ppb)	Emission counts
0	25
0.005	918
0.010	1809
0.025	4987
0.050	10143
0.075	15203
0.100	19357
0.200	39362
0.300	48587
0.400	55072
0.500	61093
0	62
0	115
0	89
0	10
0	53
0	121
0	105
0.005	928
0.005	881
0.005	985
0.005	1002
0.005	993
0.005	911
0.005	873

and compared to the calibration curve. Results of the analyses follow:

Calculate the concentration of silver in each sample. Are there any problems? How will you address each of the identified problems?

Standards Ag (ppb)	**Absorbance**
0.00	0.000
2.00	0.013
10.0	0.076
20.0	0.162
30.0	0.245
50.0	0.405
75.0	0.598
100.0	0.748

Samples	**Absorbance**
08002	0.079
08003	0.058
08004	0.129
08005	0.082
08006	0.797
08007	0.063
08008	0.009
08009	0.028
08010	0.136

36. Identify the data you would need to gather to establish the linear range for a particular analysis.

Quality Assurance Measures

Learning Objectives

Students should be able to:

Content

- Calculate the spike recovery and relative deviation between replicate samples.
- Use spike recovery and relative deviation between replicate samples to assess and validate sample analysis quality.
- Use spike recovery and relative deviation between replicate samples to assure accuracy and precision of an analysis.

Process

- Validate data. (Problem solving)
- Interpret tabulated data. (Information processing)

Prior knowledge

- Evaluation of calibration curves including linear range and limit of detection, linear regression, correlation coefficient, accuracy and precision.

Further Reading

- Harris, D.C. 2007. *Quantitative Chemical Analysis,* 7th Edition, pp.78-87. New York: W.H. Freeman.
- Skoog, D.A., F.J. Holler, and S.R. Crouch. 2007. *Principles of Instrumental Analysis,* 6th Edition, p11-22. Belmont, CA: Thomson Brooks/Cole.

Authors

Anne Falke and Shirley Fischer-Drowos

Consider this...

Consider the following analysis of silver in drinking water samples. To each of three digestion flasks, 200 mL aliquots of a drinking water sample were added. Silver was extracted by a co-precipitation procedure and collected by filtration. The precipitate was digested in 200 µl hot concentrated nitric acid and diluted to 5 mL in a dilute phosphate solution. Two of the samples were treated identically. They are called "sample" and "duplicate" in the list below. Five µL of a 10.00 ppm Ag solution was added to the third sample resulting in a 10.00 ppb spike. Standards were prepared according to the following table. The samples were analyzed by graphite furnace AAS. Results of the calibration and analysis follow:

Table 1

Run number	Ag (ppb)	Absorbance
1	Calibration Standards Calibration Blank	0.000
2	2.00	0.013
3	10.0	0.076
4	20.0	0.162
5	30.0	0.245
6	50.0	0.405
7	75.0	0.598
8	100.0	0.748
9	Instrument Performance Check (30.0 ppb)	0.243
10	Calibration Blank	0.004
11	Sample	0.030
12	Duplicate Sample	0.032
13	Laboratory Fortified Sample Marix (10.0 ppb Ag added to sample)	0.109
14	Instrument Performance Check (30.0 ppb)	0.232
15	Calibration Blank	0.003

Note that there are several analyses performed in addition to the calibration curve and sample. The additional analyses are Quality Assurance (QA) samples. These samples must be analyzed to assure the precision and accuracy of the analysis. The frequency of QA samples in an instrumental analysis is often specified by various regulatory agencies such as EPA and FDA to assure the integrity of analysis and is dependent on who is requesting the analysis.

Key Questions

1. What analyses have been performed in addition to the calibration curve? Discuss these and decide as a group why you think each of these analyses was performed. Provide an explanation for each additional analysis.

2. Plot the calibration curve, insert a trendline and determine the equation for the line.

An Instrument Performance Check is one of the intermediate calibration standards. The value obtained must be converted to a concentration using the calibration data and then compared to the expected value. Generally, recovery of an Instrument Performance Check must be between 90 and 110% (some industries apply tighter constraints) before continuing with the analysis. If recovery falls outside this range, it means that the instrumental measurements are not under control.

3. Use the calibration curve to calculate the concentration of the Instrument Performance Check.

4. Compare that value to the expected concentration (30.0 ppb.) Calculate the percent recovery (compared to the expected concentration.)

5. Does analysis of the Instrument Performance Check provide information about accuracy or precision or both? Explain.

6. As a group, decide if the recovery of the Instrument Performance Check is acceptable. As if your group were reporting this information to your client, explain the rationale for your decision using complete sentences.

The Calibration Blank is analyzed to assure that there is no instrumental contamination. It is prepared identically to the other standards (i.e. same solvents, same process), but with no analyte added. The calculated concentration must be below the Minimum Detection Limit (MDL) to continue.

7. Brainstorm some potential sources of contamination.

8. If the MDL for this method is 0.1 ppb, is there evidence of any silver contamination in the instrument? Explain your reasoning.

Good Laboratory Practice (GLP) requires that an Instrument Performance Check and a Calibration Blank be analyzed after analysis of the calibration standards, but before any samples are analyzed, regularly throughout the analysis of samples (every 10 samples is common), and after analysis of all samples. If either the check or zero standard is out of control, protocol requires that the analyst redo the calibration curve and reanalyzes all samples analyzed between the previous check standard and the one deemed unacceptable.

9. Now look at the sample labeled "Duplicate Sample." The Duplicate Sample is identical to the sample in every way except that it is prepared and analyzed separately. Does the Duplicate Sample provide information about precision, accuracy, or both? Explain in complete sentences.

The frequency at which Duplicate Sample must be prepared and analyzed is designated by the regulations governing the particular industry. Generally, an analyst prepares one duplicate for every batch of 10 or 20 samples as specified. Relative deviation is defined as the difference between the calculated concentration of the original sample and the concentration of the duplicate sample divided by the original sample concentration. Multiply by 100 to get percent relative deviation.

$$\frac{|\,\textit{Sample concentration} - \textit{duplicate concentration}\,|}{\textit{sample concentration}} \times 100\%$$

Generally, a percent relative deviation of less than 10% is acceptable. A deviation greater than 10% must be flagged and explained.

10. Calculate the concentration of the Sample and the Duplicate Sample. Calculate the percent relative deviation for this analysis. Is this acceptable? What does it tell you about the precision of this analytical technique with respect to silver?

11. Now, examine the Laboratory Fortified Sample Matrix (LFM). The LFM is prepared by adding a known amount of analyte to a particular sample. The LFM is analyzed to determine how much of the added analyte is recovered. Does this give you information about accuracy, precision, or both? Discuss this with your team members and explain in grammatically correct sentences.

12. A 10.0 ppb spike is added to the LFM. What is the concentration of silver detected in the LFM?

13. Accounting for the amount of silver detected in the sample, how much of the silver detected in the Laboratory Fortified Sample Matrix is due to the added silver?

14. Given what you know about the concentration of added silver, what percent of the added silver was detected? Explain how you determined the percent of added silver detected.

15. As a group, develop an equation for calculating percent recovery of an LFM in a similar situation with a given amount of analyte added to a sample with a known amount of the analyte.

Industry standards generally require that the recovery of the added analyte in a Laboratory Fortified Sample Matrix range from 75 – 110%. (Some industries and regulatory agencies set tighter conditions.)

16. Is the recovery of the added silver in the LFM acceptable? Explain.

17. Discuss the acceptable range for recovery of added analyte in the Laboratory Fortified Sample Matrix with your group. Why do you think there is more leeway on the lower end (75%) than on the upper end (110%)?

18. Complete the following table with the calculations for each QA parameter.

Parameter	Equation
Check Standard Recovery	
Percent Relative Deviation	
Spike Recovery	

19. As a group summarize what you learned in this activity.

a. Write down at least three key concepts.

b. Write at least one question that is lingering after completing this activity.

20. Look back at the table at the beginning of this activity. What features make it easy to make sense of the table? What features of the table would you change to make it easier to read? Do you think the table is in an appropriate format for a professional report or publication? Why or why not?

Applications

21. Analysis of lead by emission spectroscopy yields the following for the calibration standards, a sample, and the same sample spiked with 0.050 ppb lead. Calculate recovery of added lead in the Laboratory Fortified Sample Matrix.

[Pb] (ppb)	Emission counts
0	13
0.005	1018
0.010	2009
0.025	4987
0.050	10643
0.075	15403
0.100	19457
0.200	38062
Sample	1530
Laboratory Fortified Sample Matrix	9451

22. You are charged with analyzing 25 water samples for lead. EPA specifies action if more than 10% of the samples test at greater than 0.015 mg/L. Additionally, EPA mandates a QA protocol of one Sample Duplicate for every 10 samples and one Laboratory Fortified Sample Matrix for every 20 samples. The instrumental protocol calls for a calibration curve of at least four non-zero standards plus a Calibration Blank. The calibration curve must be verified before analyzing any samples, after every 10 analyses, and again at the end of the analysis. Prepare a table that lists the sequence in which the samples, spikes, duplicates and standards should be run in order to be compliant with the EPA regulations.

23. In addition to all of the quality assurance samples discussed in the activity, many regulatory agencies require analysis of a Standard Reference Material (SRM). SRMs are real samples that are well characterized. Analyte concentrations are known with a great deal of certainty. SRMs are available in almost any imaginable sample matrix (sea water, river water, estuarine water, bituminous and anthracite coal, various types of sediment, etc.). The National Institute of Standards and Technology (NIST) is a provider of SRMs. Analysis of a NIST drinking water standard certified at 19.73 + 0.02ppb silver gave the following results based on the calibration curve established from the data in Table 1.

Run #	Absorption
1	0.147
2	0.152
3	0.143
4	0.152
5	0.149
6	0.145

Based on what you have discovered in this activity, discuss whether these results are acceptable. Support your answer. If you were in the position of analyzing samples for silver concentration for a client, what would you tell that client? Be as thorough as possible.

24. In industry, time is money. Each analysis requires a specified amount of time. If each analytical run requires 2 minutes, what is the minimum amount of time required to sufficiently analyze a single sample? Consider all parts of the analysis from calibration to Quality Assurance. Explain where all of the time is used.

25. A few individuals in a small town thought there might be toxic concentrations of mercury in the drinking water. They filled an empty water bottle with water from the tap and brought it in to the local university to be analyzed. When the results came back, the concentration reported did indicate toxicity. The individuals decided to sue the town for knowingly poisoning the residents of the town. You are now employed by the town to handle the problem. What are you going to tell the lawyers? Consider everything covered in this activity.

Instrumental Calibration: Method of Standard Additions

Learning Objectives

Students should be able to:

Content

- Recognize when the method of standard additions is needed. (matrix effects)
- Distinguish between the method of standard additions and normal calibration using standards.
- Determine the concentration of an unknown in an original sample using the method of standard additions.

Process

- Interpret calibration graphs. (Information Processing)
- Calculate concentrations of unknown solutions from standard addition data. (Information Processing)
- Read and interpret a standard method. (Information Processing)

Prior knowledge

- Calibration using standards. (external and internal calibration)
- Use of statistical tests for outliers and comparing means.
- Linear regression, linearity, detection limit.
- Use of Microsoft Excel to plot data and obtain best fit lines or polynomials.
- Completion of *Quality Assurance Measures* activity.

Further Reading

- Creed, J.T., Martin, T.D., and O'Dell, J.W. Method 200.9, *Determination of Trace Elements by Stabilized Temperature Graphite Furnace Atomic Absorption*, Revision 2.2. US Environmenta Progection Agency [online] Nov. 4, 2010. http://water.epa.gov/scitech/methods/cwa/bioindicators/upload/2007_07_10_methods_method_200_9.pdf
- Harris, D.C., 2010. *Quantitative Chemical Analysis,* 8th Edition, W.H. Freeman: USA, p.106-108, 330, and 494.
- Harvey, D.J. *Chem.* Ed. 2002, 79, pp.613-615.

Authors

Christine Dalton and Mary Walczak

Section 1. Determining When the Method of Standard Additions is Needed

In this section we will perform a number of different quality assurance checks. For convenience, Appendix C contains an abbreviated list of the checks that will be conducted. You may wish to tear off Appendix C for use throughout the activity.

Consider this...

Determining the Linear Dynamic Range

The following data tables were collected for analysis of industrial waste samples for lead using a graphite furnace atomic absorption instrument according to EPA Method 200.9 *(Determination of Trace Elements by Stabilized Temperature Graphite Furnace Atomic Absorption)*. Table 1 shows the data for five solutions of lead at the same concentration (7.5 ppb) analyzed to demonstrate instrument stability. The % Relative Standard Deviation (RSD) must be less than 5% for the instrument stability check to pass. *Note: Appendix A includes an excerpt from Section 3 of the EPA Method with definitions of useful terms.*

Table 1 *Solutions Prepared for Atomic Absorption Analysis of Industrial Waste Samples for Lead*

Demonstrate Instrument Stability (Section 11.4.3*)		
Solution Number	**Concentration Pb Standard (ppb)**	**Absorbance**
1	7.5	0.036
2	7.5	0.036
3	7.5	0.035
4	7.5	0.037
5	7.5	0.034

* The section number refers to the EPA Method 200.9

Key Questions

1. Solutions 1–5 are analyzed to demonstrate instrument stability. According to the method, "the resulting relative standard deviation (RSD) of absorbance signals must be <5%." What is the %RSD for these five solutions? Does the instrument pass the stability check? Indicate your decision on the Table in Appendix C. *(Note: We will collect all our quality assurance checks in one place in Appendix C.)*

2. The method says "instrument stability must be demonstrated by analyzing a standard solution with a concentration 20 times the Instrument Detection Limit (IDL) a minimum of five times." Solutions 1-5 are run to demonstrate instrument stability. What is the IDL for this instrument? Check your answer by conferring with another group. Resolve any discrepancies before continuing.

Consider this...

Table 2 below shows the calibration run for lead following EPA method 200.9 and the solutions used to determine the method detection limit (MDL). The plot for the data in Table 2 is shown on the next page along with the least-squares line and correlation coefficient (R^2) for different concentration ranges.

Table 2 *Solutions Prepared for Atomic Absorption Analysis of Industrial Waste Samples for Lead*

Calibration Run (Section 11.4.4)		
Solution Number	**Concentration Pb Standard (ppb)**	**Absorbance**
6	0	0.001
7	1.6	0.018
8	8	0.048
9	20	0.085
10	40	0 186
12	60	0.276
13	90	0.422
14	110	0.535
15	140	0.657
16	170	0.704
17	200	0.795
Method Detection Limit (Section 9.2.4)		
Solution Number	**Concentration Pb Standard (ppb)**	**Absorbance**
18	1.0	0.001
19	1.0	0.002
20	1.0	0.001
21	1.0	0.003
22	1.0	0.002
23	1.0	0.002
24	1.0	0.002

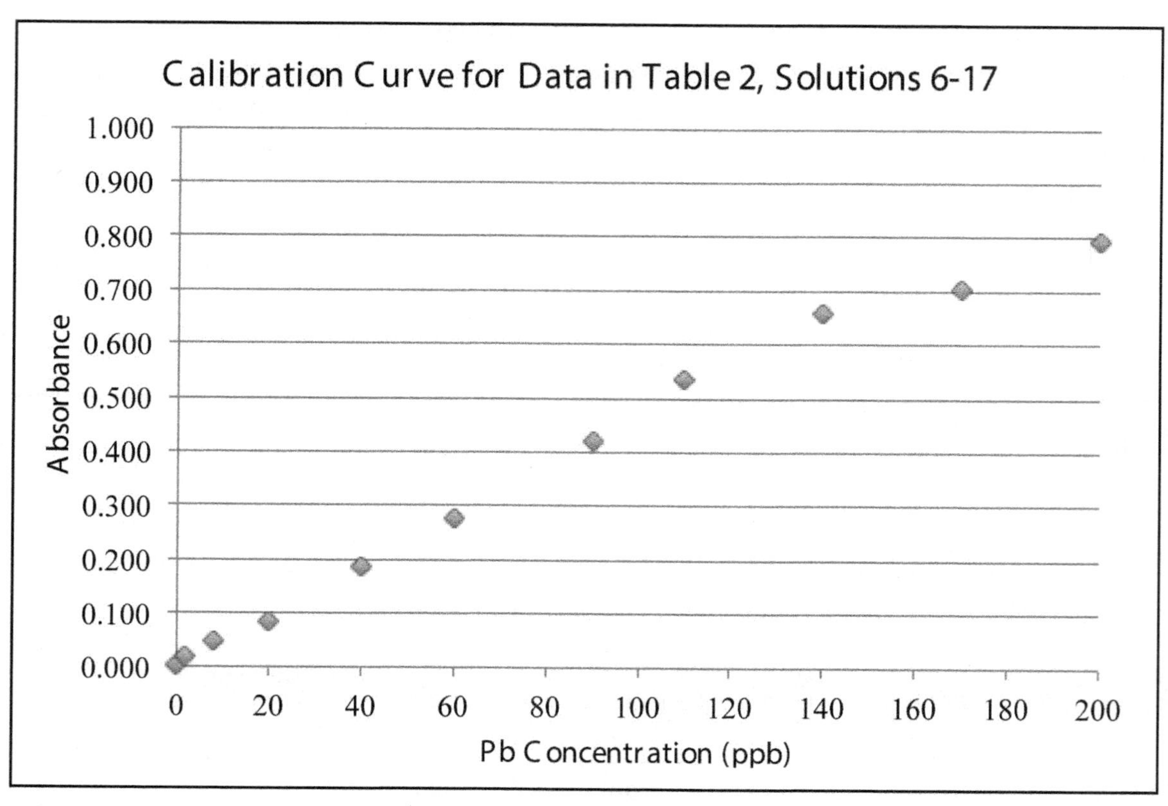

range	equation	R^2
0 to 110 ppb	y = 0.00474 x + 0.001	0.9979
0 to 140 ppb	y = 0.00471 x + 0.002	0.9988
0 to 170 ppb	y = 0.00440 x + 0.012	0.9907
0 to 200 ppb	y = 0.00416 x + 0.021	0.9868

Key Questions

3. What is the upper limit for Pb analysis using this method? How did you determine this limit?

4. The method detection limit (MDL) is determined once per year by "processing seven replicate samples through the entire method." These seven replicates are prepared "using reagent water (blank) fortified at a concentration of two to three times the estimated [IDL]." To calculate the MDL, the Student's t value for a confidence level of 99% and for n-1 degrees of freedom is multiplied by the standard deviation of the concentrations of the seven replicates. If t=3.14 for seven replicates, what is the MDL for this analysis using solutions 18-24 in Table 2? (Hint: We want the MDL in concentration units (ppb), so we must first change the absorbance values for solutions 18-24 into concentrations.)

5. Using the information in Q2-Q4, fill n the following table

IDL	
MDL	
Upper limit	

The linear dynamic range for an analysis is the lowest to highest concentrations that can be measured with the method. The lower limit of quantitation (LLOQ) is the smallest concentration that can be measured with reasonable accuracy. Based on the equations below, where s is the standard deviation of the concentrations of the solutions used to determine the MDL and m is the slope of the best calibration curve, the MDL can be used to calculate the LLOQ:

$$\text{MDL} = \frac{3 * s}{m} \qquad \text{LLOQ} = \frac{10 * s}{m}$$

$$\frac{s}{m} = \frac{\text{MDL}}{3} \qquad \text{LLOQ} = \frac{10 * s}{m} = \frac{10 * \text{MDL}}{3}$$

Using the information entered in the table above, determine the "linear dynamic range" for Pb analysis using this method. Compare your answer to that of another group.

Linear dynamic range	

Consider this...

Assessing Laboratory Performance

According to the EPA 200.9 method, assessing laboratory performance is mandatory on a daily basis. If any of the Quality Controls do not pass, then the sample data is invalid and the problem must be found and corrected prior to running samples. Solutions 25 through 48 in Table 3 include several solutions that are utilized for assessing laboratory performance along with analysis of several samples.

Table 3 *Solutions Prepared for Atomic Absorption Analysis of Industrial Waste Samples for Lead*

Sample analysis (11.4) with Performance Assessment (9.3 and 9.4)		
Solution Number	**Sample Name**	**Absorbance**
25	Laboratory Reagent Blank	0.004
26	Calibration Blank	0.002
27	Instrument Performance Check (25 ppb)	0.123
28	Laboratory Fortified Blank (25 ppb)	0.122
29	Sample 1	0.408
30	Sample 2	0.521
31	Sample 3	0.446
32	Duplicate Sample 3	0.448
33	Sample 4	0.512
34	Sample 5	0.926
35	Sample 6	0.489
36	Laboratory Fortified Sample Matrix: Sample 6 + 25 ppb Spike	0.568
37	Sample 7	0.506
38	Sample8	0.579
39	Laboratory Reagent Blank	0.003
40	Calibration Blank	0.001
41	Instrument Performance Check (25 ppb)	0.112
42	Sample 9	0.482
43	Sample 10	0.009
44	Sample 11	0.819
45	Sample 12	0.486
46	Laboratory Reagent Blank	0.002
47	Calibration Blank	0.002
48	Instrument Performance Check (25 ppb)	0.125

Key Questions

Exploring the Solutions from Table 3

6. Referring to solutions numbered 1-48 in Tables 1-3, match all 48 solution numbers with the eight categories of solutions listed in column one of the table. Add the solution numbers to the second column of the table from Appendix C.

Solutions	Solutions from Tables 1-3
Instrument Stability Check	
Calibration Standards	
Method Detection Limit	
Laboratory Reagent Blank	
Calibration Blank	
Instrument Performance Check	
Laboratory Fortified Blank	
Samples	
Duplicate Samples	
Laboratory Fortified Sample	

Assessing blanks for contamination

7. According to the EPA method there are several types of "blanks" used in the analytical run. Solutions 25-48 in Table 3 include several of these blank solutions. Identify the different types of blanks in Table 3 and add these to the following table. Specify the solution numbers for each type from Table 3 and add those to the following table. List the purpose and composition of each type of blank in the run. Note: Appendix A includes an excerpt from Section 3 of the EPA Method with definitions of useful terms, including several types of blanks.

Solution	Solution Number(s)	Purpose of Solution	Composition of Solution	Requirement
			Contains all reagents in the same volumes as used in processing samples	[Analyte] < 2.2 x MDL
		Used to check that the instrument is zeroed correctly throughout the run		[Analyte] < MDL
	28			% Rec = 85-115%

8. Using the best least-squares line (determined in Q3) calculate the concentration for each laboratory reagent blank and calibration blank. Add the concentrations of these blanks to the last column of Table 3 in Appendix B. *Note: Do not add any other concentrations to this table until you are told to do so.*

9. Using the information in the table in Q7, determine if each laboratory reagent blank (LRB) and calibration blank in Table 3 meets the requirement for that type of blank to pass the contamination check. Recall that the MDL and IDL were determined in Q5. Do all of the tests pass for the LRBs and calibration blanks? If all of the LRBs and calibration blanks pass the test, then add a check in the "Pass?" column in the table in Appendix C. If they do not pass the test, then write "Fail" in the "Pass?" column.

Exploring samples and assessing sample duplicates

10. Table 3 includes several solutions labeled as "samples." How many different samples were analyzed? How many replicates of the samples were performed? According to the EPA method, a replicate must be analyzed for every 10 samples. Does this run meet this requirement?

11. What are the differences between solutions 31 and 32? What is the purpose for running both of these samples? If the method states that the difference between a sample and a sample duplicate must be less than 10%, does this run pass this criterion? Add a check or "Fail" to the table in Appendix C to designate if the duplicate passes the criterion for performance assessment.

Checking concentrations of known solutions

12. Calculate the concentration of the Laboratory Fortified Blank (LFB) in Table 3 and add the concentration to Appendix B. What is the percent recovery (i.e., calculated value/expected value x 100) for the LFB if the amount of added Pb is 25 ppb? Does this value fall within the acceptable limits? Add a check or "Fail" in the "Pass?" column in the table in Appendix C to specify passing or failing of the LFB.

13. The run sequence includes several Instrument Performance Checks as shown in Q7. What is the purpose of including these samples in the run?

14. Calculate the concentration of the instrument performance check (IPC) in Table 3 and add the concentration for each check to the table in Appendix B. The instrument performance check is done using a certified reference material with a known concentration of 25 ppb. Based on the calculated concentration of the standard reference material, determine the percent recovery of the reference material (i.e., calculated value/expected value x 100). The EPA method states that the first IPC check should fall within 5% of the calibration, while later IPC checks can be within 10% of the calibration. Do the IPC check concentrations fall within the acceptable limits? Add a check or "Fail" in the "Pass?" column in the table in Appendix C to indicate if the IPC passes the performance check.

15. What are the differences between solutions 35 and 36? What is the purpose for running both of these samples? The EPA method states that the Laboratory Fortified Sample Matrix must be analyzed once for every ten samples, and it must be a duplicate of the aliquot used for sample analysis. Does sample 36 meet this criterion?

16. Calculate the concentrations of solutions 35 and 36 and add them to Table 3. For the Laboratory Fortified Sample Matrix to pass the performance assessment, the percent recovery of the analyte must be within 70-130%. This is calculated by subtracting the unfortified sample's concentration (solution 35) from the fortified sample's concentration (solution 36) and dividing by the concentration of analyte added to the sample and multiplying the result by 100. Does sample 36 pass the performance check? Add a check or "Fail" in the "Pass?" column in the table in Appendix C to indicate if the Laboratory Fortified Sample Matrix (LFM) passes the test.

17. In Q12 through Q16, solutions of known concentration were analyzed and the percent recovery was checked against a criterion of the method. What is the difference between the LFB, the IPC, and the LFM if each had 25 ppb Pb added?

18. If the LFM is the only solution of known concentration to fail the criterion for percent recovery, brainstorm reasons why this solution would not be within the acceptable range? Come to consensus.

19. Review the data you have accumulated in the Table from Appendix C. Do all parts of the assessment pass so that the analysis of the samples is acceptable? Explain how you came to your conclusion.

In chemical analysis, the presence of other analytes in the sample can interfere with the detection of the analyte of interest. This is especially true of natural samples, such as pond water. When this occurs it is called a "matrix effect" because the compounds in the sample interfere with analysis. In such cases, the method of standard additions can be used to eliminate differences between samples and calibration standard solutions.

20. What is the logical next step in this analysis?

Section 2. Determining Concentration from Standard Addition Data

Consider this...

Calibration standards were prepared using 100-mL volumetric flasks and a 75.0 ppm standard solution of analyte. The amount of standard added to each flask is listed in Table 4. Each flask was then diluted to volume with deionized water and thoroughly mixed. *Note that solution shading (see legend) is not meant to indicate the lack of mixing but just illustrate relative amounts.*

Table 4 ***Calibration Standards***

Solution	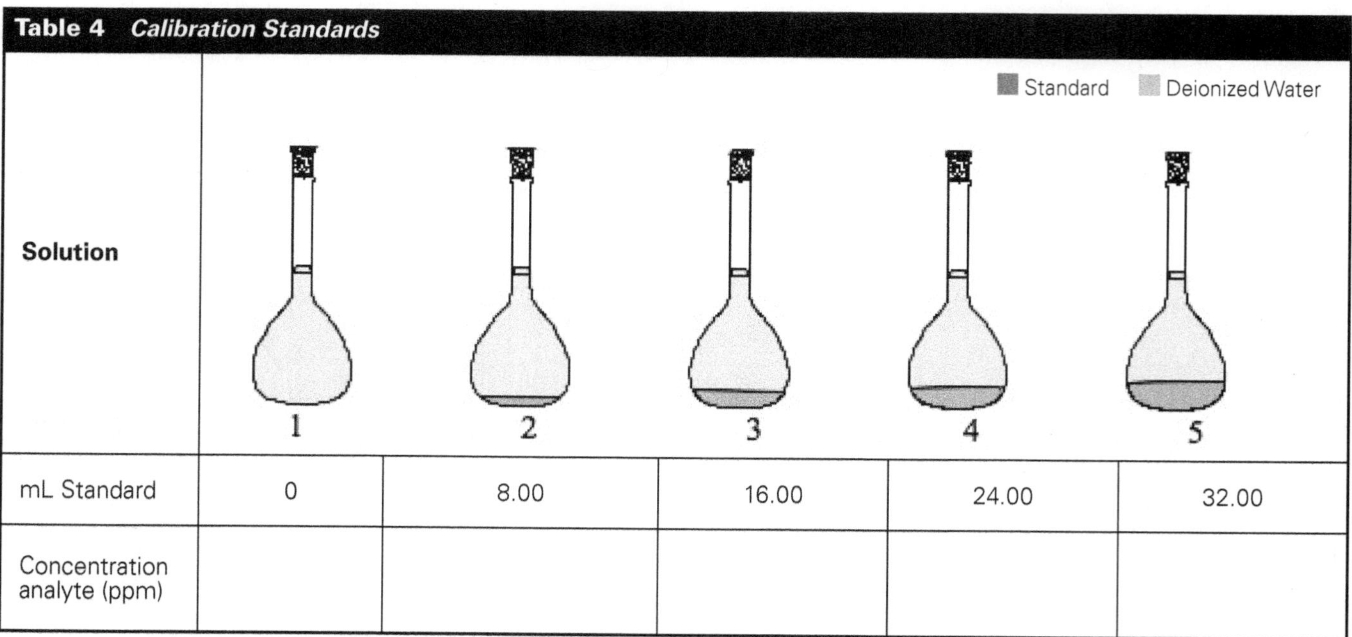				
mL Standard	0	8.00	16.00	24.00	32.00
Concentration analyte (ppm)					

Key Questions

21. What happens to the relative amounts of standard and water among the calibration standards?

22. As a group, consider the calibration standards samples in Table 4. First, devise a general method to calculate the concentration of analyte in a prepared solution.

Then divide solutions 1-5 among group members and calculate the concentrations. Add the answers to the Table and check that concentrations increase with the amount of standard used.

23. Suppose you wanted to make a 16.0 ppm calibration standard. Describe how you would prepare this solution.

Consider this...

A set of Standard Addition solutions were prepared using 100-mL volumetric flasks, a 120. ppm standard solution of analyte and a solution containing an unknown amount of analyte. Ten mL of the unknown solution were added to each flask using a volumetric pipet. Increasing amounts of the 120. ppm standard were also added to the flasks, as illustrated in Table 5. Each flask was then diluted to volume with deionized water and thoroughly mixed. *Note that solution shading (see legend) is not meant to indicate the lack of mixing but just illustrate relative amounts.*

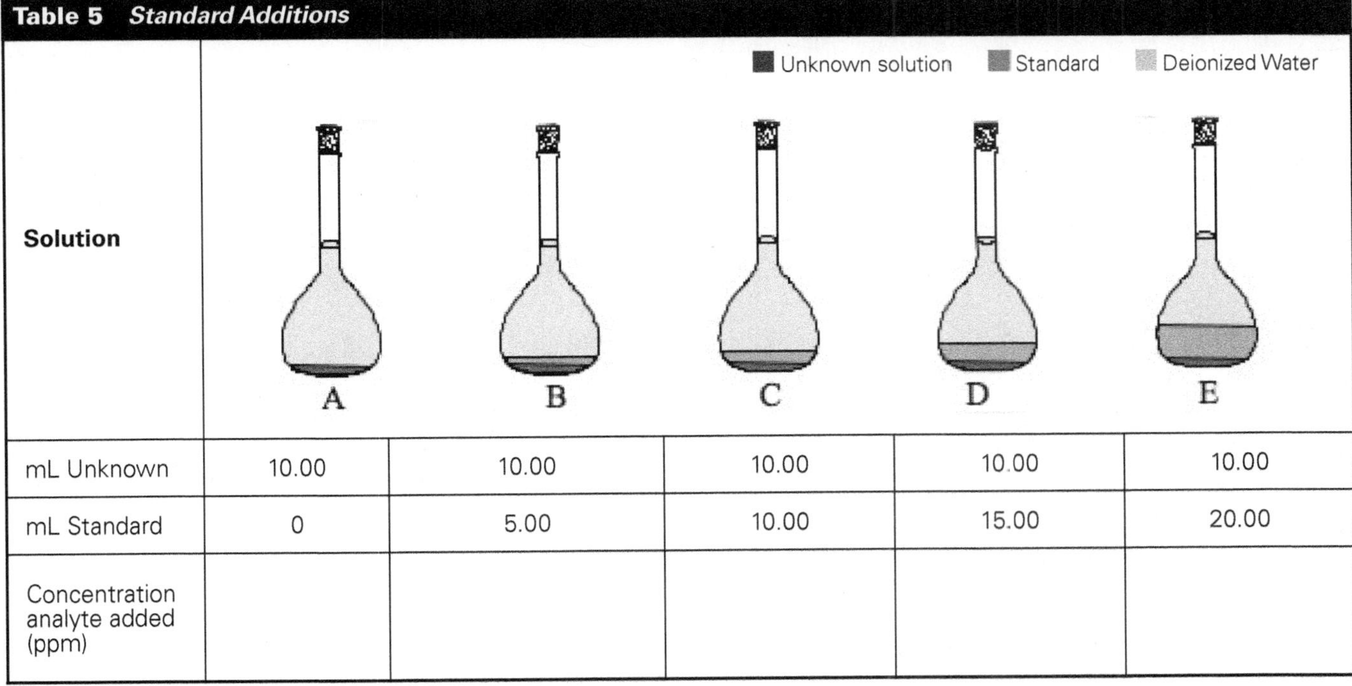

Table 5 ***Standard Additions***

Solution	A	B	C	D	E
mL Unknown	10.00	10.00	10.00	10.00	10.00
mL Standard	0	5.00	10.00	15.00	20.00
Concentration analyte added (ppm)					

24. Consider the relative amounts of the three solution components. What happens to the amounts among solutions A-E?

Unknown Solution	Increases	Decreases	Stays the Same
Standard	Increases	Decreases	Stays the Same
Water	Increases	Decreases	Stays the Same

25. As a group, consider the standard addition samples in Table 5. What are the concentrations of analyte *added* to each solution? Fill in the answers in the Table. Check your answers by comparing with another group.

26. How would preparing a 16.0 ppm standard addition sample be different from the standard prepared in Q23?

Consider this...

The atomic absorbance of each solution in Tables 4 and 5 was measured. The resulting graphs are shown below.

Figure 1 *Absorbance data for solutions from (a) Table 4 and (b) Table 5*

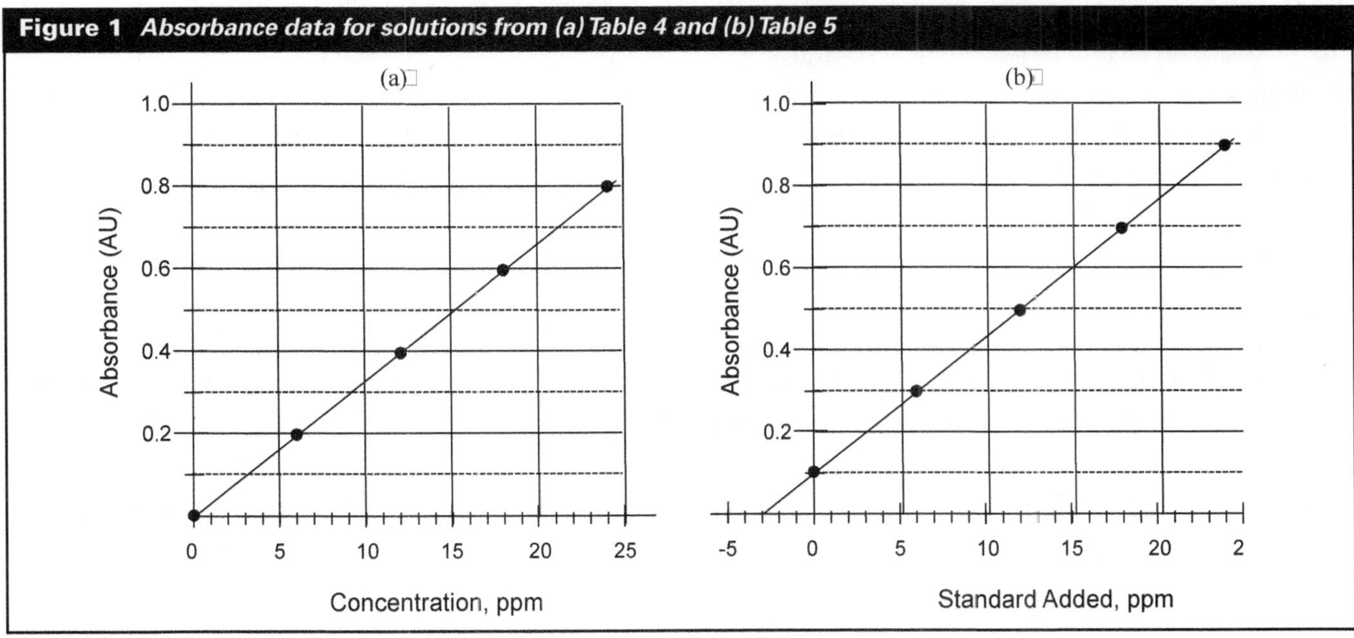

Key Questions

27. Examine Figure 1 carefully. Which graph (a) or (b) corresponds to analysis of the *Standards?* Which corresponds to the *Standard Addition* solutions?

28. Add the absorbances for each solution to Tables 4 and 5, reproduced in part below. Estimate the absorbance values by looking at the graphs in Figure 1.

Table 4 *Calibration Standards*

Solution	1	2	3	4	5
mL Standard	0	8.00	16.00	24.00	32.00
Concentration analyte (ppm)	0	6	12	18	24
Absorbance					

Table 5 *Standard Additions*

Solution	A	B	C	D	E
mL Unknown	10.00	10.00	10.00	10.00	10.00
mL Standard	0	5.00	10.00	15.00	20.00
Concentration analyte added (ppm)	0	6	12	18	24
Absorbance					

29. Based on the absorbances of the five solutions in each graph, which solution has the higher concentration of analyte for the same concentration value on the x-axis?

30. If the absorbance of an unknown solution of the analyte is measured, the concentration of that solution can be determined from the calibration standards. If an unknown solution has an absorbance of 0.50, what is the concentration of analyte in the unknown solution?

31. Consider the 12 ppm solution. The absorbances are different for the calibration standard solution and for the standard addition solution containing 12 ppm analyte.

a. By how much is the absorbance of the 12 ppm standard addition solution greater than the 12 ppm calibration standard solution?

b. Examine the absorbances of the other solutions in the calibration standards and standard addition datasets. How much greater is the absorbance for these other solutions?

c. Where does the "extra" analyte come from?

32. What is the absorbance of the standard addition solution containing no added standard in Figure 1b?

33. Use the calibration graph in Figure 1a to determine the concentration of analyte in a solution with this absorbance.

34. Circle the concentration (x-value) that corresponds to zero absorbance on the standard addition curve (Figure 1b).

35. Compare your answers to Q33 and Q34. What is the mathematical relationship between these two results?

In laboratory situations where standard addition is needed it is unlikely that the slopes of the two plots in figure 1 would have the same value. In this activity the same slopes were used to illustrate the mathematical relationship required to analyze standard addition data.

36. Ordinarily, data for either standards *or* standard addition are collected. If you had only standard addition data, explain how to find the *concentration* of an unknown analyte from standard addition data.

37. You are training a new laboratory technician. How do you teach them about when standard addition methods are necessary?

Applications

38. Calculate the concentration for each sample and duplicate in Table 3.

(a) Is the concentration of any sample below the MDL? If so, what would you report as the result for that sample?

(b) Is the concentration of any sample above the upper limit of the dynamic range? If so, what should be done for that sample?

39. The EPA method used as the basis of this activity says that the MDL is determined by "processing seven replicate samples through the entire method." These seven replicates are prepared "using reagent water (blank) fortified at a concentration of two to three times the estimated [IDL]." The MDL is calculated, according to the EPA method, by multiplying the standard deviation for the seven replicates by the Student's t value for a confidence level of 99% and for n-1 degrees of freedom. The Harris text specifies that the LOD (=MDL) is found by 3*s/m, where s is the standard deviation of $n \geq 7$ replicates and m is the slope of the calibration line. Comment on the difference between these two methods of determining the MDL.

40. For each experimental consideration listed in the Table below, circle the method (Standards or Standard Addition) that is better in your opinion. Give a reason for your choice.

Experiment Consideration			**Reason**
# unknowns determined per set of solutions	Standards	Standard Addition	
Time required to prepare solutions	Standards	Standard Addition	
Accuracy obtained when a matrix effect exists	Standards	Standard Addition	
Cost of materials	Standards	Standard Addition	
Cost of labor	Standards	Standard Addition	

41. Considering all the factors in Q40, which method (standards or standard addition) is better under what circumstances? *Make a recommendation for your supervisor regarding future analyses done in your lab.*

42. Suppose you wanted to determine the amount of calcium in milk. Which approach is easier: to mimic the milk matrix in your standards or to add standard to milk and use the existing matrix?

43. Consider the shape of most atomic absorption calibration curves (see examples below). Why is it important to use solutions that have concentrations in the working range?

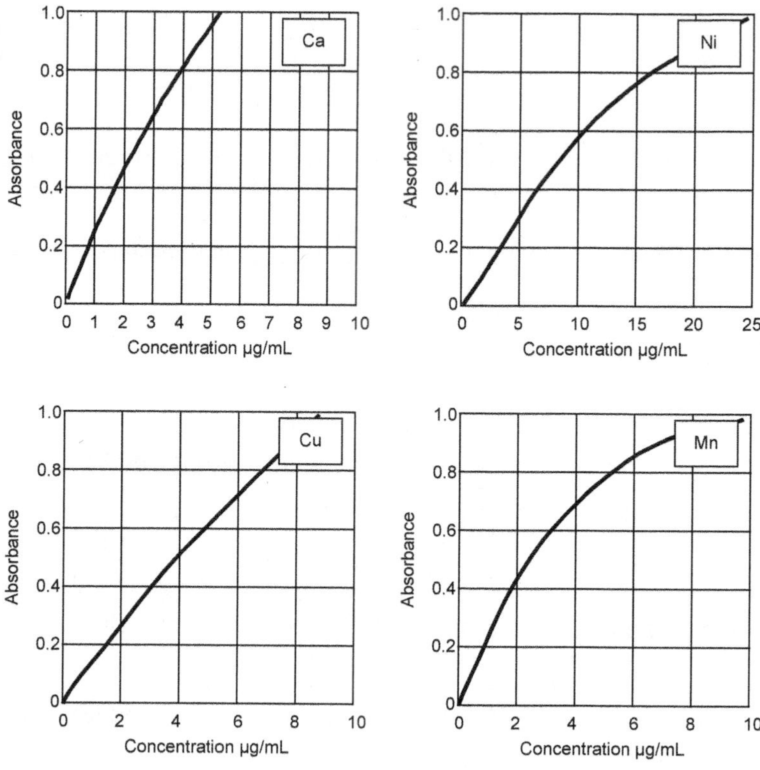

44. Sketch the absorbance vs. concentration graph expected when the standard addition method is used in visible spectroscopy. Explain how to find the concentration of the unknown from this graph.

45. (a) The calibration graph at the right was obtained by AAS using solutions containing known amounts of Zn^{2+}. An unknown sample was measured under the same conditions and the absorbance was 0.560. Determine the Zn^{2+} concentration in the unknown sample.

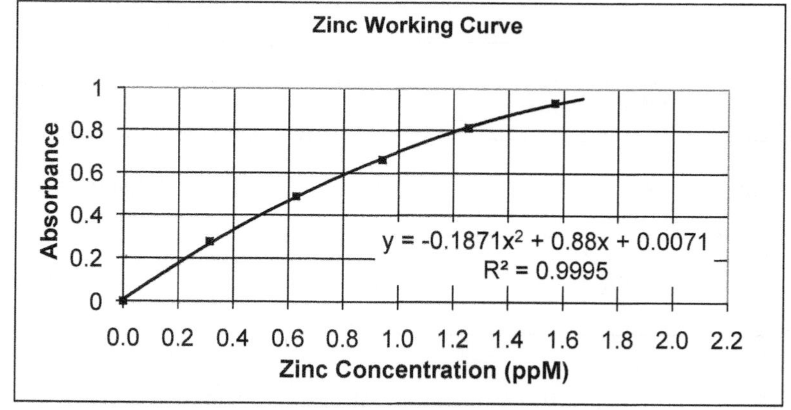

(b) Explain the difference between the method used to determine Zn in problem (a) and the standard addition method. How would you prepare the Zn solutions if you wanted to use the standard addition method instead? Be specific.

(c) The Zn-containing unknown in problem (a) was diluted prior to analysis in three serial dilutions: 10.00 mL of the original solution was diluted to 100.00 mL; 20.00 mL of the resulting solution was diluted to 100.00 mL and 20.00 mL of that solution diluted to 250.00 mL. What was the Zn concentration in the original solution?

46. A standard addition analysis of a Pb-containing solution was done using AAS. The solutions were prepared by adding 15.00 mL of a diluted unknown Pb solution to a varying amount of 505.8 ppm Pb stock solution, as depicted in the table below. Note that deionized water was NOT added to these solutions to give a constant total volume.

Sample	Diluted Unknown Solution (mL)	505.8 ppm Pb standard (mL)	Absorbance
1	15.00	0.00	0.019
2	15.00	0.09	0.181
3	15.00	0.29	0.440
4	15.00	0.59	0.663
5	15.00	0.81	0.749
6	15.00	1.00	0.798

(a) Plot the absorbance of the solution as a function of ppm Pb added and determine the concentration of Pb in the diluted unknown solution. What is the uncertainty in this Pb concentration?

(b) The unknown solution was prepared by serial dilution as follows: 10.00 mL of the original unknown solution was diluted to 100.00 mL. The resulting solution was diluted serially 10.00 mL to 100.00 mL three more times. What is the concentration of Pb in the original unknown solution?

(c) Based on the uncertainty in the unknown Pb solution (part a) and the tolerances of the volumetric glassware used in diluting the unknown Pb solution from the original concentration, determine the uncertainty in the concentration of the original unknown solution.

47. Lake phosphorous levels were determined spectrophotometrically using a molybdenum assay (EPA Monitoring and Assessment: Phosphorous, http://water.epa.gov/type/rsl/ monitoring/vms56.cfm). The method of standard additions was employed. The samples were prepared in 25-mL volumetric flasks using lake water and standard phosphorous solution, as illustrated in the table.

Sample	Lake Water (mL)	3.00 ppm Phosphorous standard (mL)	Absorbance
1	15.00	0.00	0.109
2	15.00	1.00	0.246
3	15.00	2.00	0.440
4	15.00	3.00	0.645
5	15.00	4.00	0.789
6	15.00	5.00	0.943

Plot the absorbance of the solution as a function of ppm P added and determine the concentration of P in the lake water sample. What is the uncertainty in this P concentration?

48. Zinc in supplements was determined and the following data obtained. Analyze the results and determine whether standard addition methods are required.

Solution Number	Concentration Zn Standard (ppm)	Absorbance
1	0.00	0.002
2	0.30	0.209
3	0.60	0.423
4	0.90	0.601
5	1.20	0.811
6	1.50	0.987
Solution Number	**Solution Description**	**Absorbance**
9	Check Standard (0.9 ppm)	0.637
10	Zero standard	0.002
11	Sample	0.548
12	Duplicate	0.572
13	Spike (0.6 ppm added)	0.922
14	Check Standard (0.9 ppm)	0.624
15	Zero standard	0.003

49. The *Standard Addition* solutions A-E were prepared by placing 10.00 mL of the unknown solution into each of five 100-mL volumetric flasks. By what factor has the unknown been diluted when the volumetric flasks are diluted to volume?

Appendix A: Glossary of Definitions

(Excerpt from EPA Method 200.9 Section 3)

Calibration Blank - A volume of reagent water acidified with the same acid matrix as in the calibration standards. The calibration blank is a zero standard and is used to auto-zero the AA instrument (Section 7.10.1).

Calibration Standard (CAL) - A solution prepared from the dilution of stock standard solutions. The CAL solutions are used to calibrate the instrument response with respect to analyte concentration (Section 7.9).

Dissolved Analyte - The concentration of analyte in an aqueous sample that will pass through a 0.45 µm membrane filter assembly prior to sample acidification (Section 11.1).

Field Reagent Blank (FRB) - An aliquot of reagent water or other blank matrix that is placed in a sample container in the laboratory and treated as a sample in all respects, including shipment to the sampling site, exposure to the sampling site conditions, storage, preservation, and all analytical procedures. The purpose of the FRB is to determine if method analytes or other interferences are present in the field environment (Section 8.5).

Instrument Detection Limit (IDL) - The concentration equivalent to the analyte signal which is equal to three times the standard deviation of a series of ten replicate measurements of the calibration blank signal at the same wavelength.

Instrument Performance Check (IPC) Solution - A solution of method analytes, used to evaluate the performance of the instrument system with respect to a defined set of method criteria (Sections 7.11 and 9.3.4).

Laboratory Duplicates (LD1 and LD2) - Two aliquots of the same sample taken in the laboratory and analyzed separately with identical procedures. Analyses of LD1 and LD2 indicates precision associated with laboratory procedures, but not with sample collection, preservation, or storage procedures.

Laboratory Fortified Blank (LFB) - An aliquot of LRB to which known quantities of the method analytes are added in the laboratory. The LFB is analyzed exactly like a sample, and its purpose is to determine whether the methodology is in control and whether the laboratory is capable of making accurate and precise measurements (Sections 7.10.3 and 9.3.2).

Laboratory Fortified Sample Matrix (LFM) - An aliquot of an environmental sample to which known quantities of the method analytes are added in the laboratory. The LFM is analyzed exactly like a sample, and its purpose is to determine whether the sample matrix contributes bias to the analytical results. The background concentrations of the analytes in the sample matrix must be determined in a separate aliquot and the measured values in the LFM corrected for background concentrations (Section 9.4).

Laboratory Reagent Blank (LRB) - An aliquot of reagent water or other blank matrices that are treated exactly as a sample including exposure to all glassware, equipment, solvents, reagents, and internal standards that are used with other samples. The LRB is used to determine if method analytes or other interferences are present in the laboratory environment, reagents, or apparatus (Sections 7.10.2 and 9.3.1).

Linear Dynamic Range (LDR) - The concentration range over which the instrument response to an analyte is linear (Section 9.2.2).

Matrix Modifier - A substance added to the graphite furnace along with the sample in order to minimize the interference effects by selective volatilization of either analyte or matrix components.

Method Detection Limit (MDL) - The minimum concentration of an analyte that can be identified, measured, and reported with 99% confidence that the analyte concentration is greater than zero (Section 9.2.4 and Table 2).

Control Sample (QCS) - A solution of method analytes of known concentrations which is used to fortify an aliquot of LRB or sample matrix. The QCS is obtained from a source external to the laboratory and different from the source of calibration standards. It is used to check either laboratory or instrument performance (Sections 7.12 and 9.2.3).

Solid Sample - For the purpose of this method, a sample taken from material classified as either soil, sediment or sludge.

Standard Addition - The addition of a known amount of analyte to the sample in order to determine the relative response of the detector to an analyte within the sample matrix. The relative response is then used to assess either an operative matrix effect or the sample analyte concentration (Sections 9.5.1 and 11.5).

Stock Standard Solution - A concentrated solution containing one or more method analytes prepared in the laboratory using assayed reference materials or purchased from a reputable commercial source (Section 7.8).

Total Recoverable Analyte - The concentration of analyte determined to be in either a solid sample or an unfiltered aqueous sample following treatment by refluxing with hot dilute mineral acid(s) as specified in the method (Sections 11.2 and 11.3).

Water Sample - For the purpose of this method, a sample taken from one of the following sources: drinking, surface, ground, storm runoff, industrial or domestic wastewater.

Appendix B

Copy of Table 3 to tear off and use for questions 6-20.

Table 3 ***Solutions Prepared for Atomic Absorption Analysis of Industrial Waste Samples for Lead***

Sample analysis (11.4) with Performance Assessment (9.3 and 9.4)			
Solution Number	**Sample Name**	**Absorbance**	**Concentration**
25	Laboratory Reagent Blank	0.004	
26	Calibration Blank	0.002	
27	Instrument Performance Check (25 ppb)	0.123	
28	Laboratory Fortified Blank (25 ppb)	0.122	
29	Sample 1	0.408	
30	Sample 2	0.521	
31	Sample 3	0.446	
32	Duplicate Sample 3	0.448	
33	Sample 4	0.512	
34	Sample 5	0.926	
35	Sample 6	0.489	
36	Laboratory Fortified Sample Matrix: Sample 6 + 25 ppb Spike	0.568	
37	Sample 7	0.506	
38	Sample 8	0.579	
39	Laboratory Reagent Blank	0.003	
40	Calibration Blank	0.001	
41	Instrument Performance Check (25 ppb)	0.112	
42	Sample 9	0.482	
43	Sample 10	0.009	
44	Sample 11	0.819	
45	Sample 12	0.486	
46	Laboratory Reagent Blank	0.002	
47	Calibration Blank	0.002	
48	Instrument Performance Check (25 ppb)	0.125	

Appendix C

Table to Complete as Quality Assurance Checks are Performed.

Solutions	Criterion for Acceptance	Solutions from Table 1-3 (Q6)	Pass?
Instrument Stability Check	<5% RSD		
Calibration Standards	N/A		N/A
Method Detection Limit	N/A		N/A
Laboratory Reagent Blank	<2.2 x MDL		
Calibration Blank	< MDL		
Samples	N/A		N/A
Replicate Samples	< 10% difference		
Laboratory Fortified Blank	% recovery=85-115%		
Instrument Performance Check	within±5%		
Laboratory Fortified Sample	% recovery=70-130%		

Interlaboratory Comparisons

Learning Objectives

Students should be able to:

Content

- Students will be able to explain how random and systematic error affects experimental results.
- Students will be able to describe what the purpose of an interlaboratory study is and how it might be conducted.
- Students will be able to explain how spike recovery may be used to evaluate accuracy of a method and of a laboratory's performance.

Process

- Students will engage in drawing inferences from data presented in graphical form. (Information processing, Critical thinking)

Prior knowledge

- Students should be able to describe accuracy and precision at least in crude terms.

Further Reading

- Bauer, C.F., S.M. Koza, and T.F. Jenkins 1990. "Collaborative Test Results for a Liquid Chromatographic Method for the Determination of Explosives Residues in Soil" J. Assoc. Off. Anal. Chem. 73: pp.541-552. Accepted as a standard method by Association of Official Analytical Chemists (991.09), ASTM (D5143-90), and EPA (SW-846 Method 8330).
- Harris, D. C. 2007. Quantitative Chemical Analysis, 7th Edition, Chapter 3-5. New York: W.H. Freeman.
- Skoog, D. A., D. M. West, F. J. Holler, and S. R. Crouch. 2004. Fundamentals of Analytical Chemistry, 8th Edition, Chapter 5-7. Belmont, CA: Thompson Brooks/Cole.

Author

Christopher F. Bauer

Consider this...

New analytical methods are usually developed in one particular laboratory by a small number of scientists. If the method is to be useful, the results need to be reliable when used by large numbers of scientists working in many different laboratories. One way of evaluating how well a method performs when applied by many users is called an *interlaboratory study.*

In an interlaboratory study, several different laboratories carry out an analysis of the same materials, and then the results are compared. If all laboratories follow the same procedure, then the comparison provides information about how well the procedure performs under realistic conditions.

The data discussed here are from an actual interlaboratory study of a method for measuring Army munitions residues in soils. The method involves extracting 2 grams of soil with 10 mL of acetonitrile (CH_3CN) for 18 hours followed by filtration and analysis of the filtrate by liquid chromatography. There are several different munitions substances present in munitions waste. Liquid chromatography is used to separate these substances. Ultraviolet absorption spectrometry is used to detect and quantify the amounts of each substance.

Two analytes will be considered: RDX and tetryl.

RDX

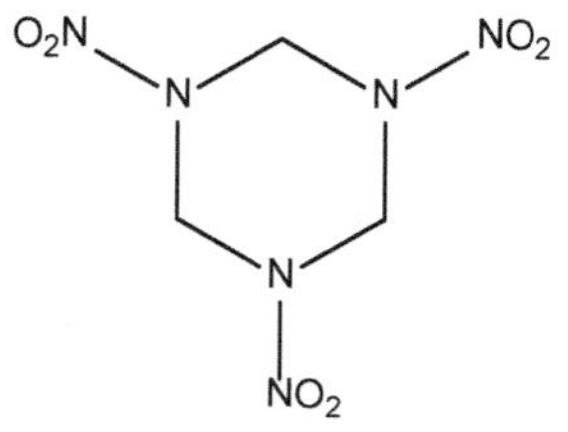

Tetryl

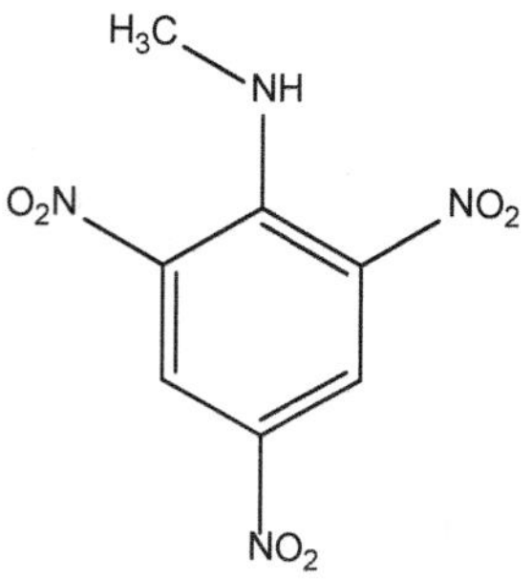

Four portions of the same clean soil were spiked with these analytes (i.e. exactly known amounts of analyte were added to the soils in solution form in a solvent that evaporates easily). The concentrations were chosen so that there would be an identical pair of high concentration spikes and an identical pair of low concentration spikes, with about 10% difference in concentration between the two. The laboratories (numbered 1-7) did not know the actual concentrations (mass of analyte per mass of soil). This is called a blind analysis. The reported results are in Tables 1 and 2.

Table 1 *Reported concentrations of RDX (µg/g)*

	Low spike 1	Low spike 2	High spike 1	High spike 2
	90.4	90.4	100.4	100.4
Lab ID	**Concentration found by using the method**			
1	87.9	85.2	98.3	96.3
2	87.4	88.2	99.7	98.9
3	87.8	83.6	99.6	94.4
4	85.7	87.2	91.6	92.7
5	102.0	102.0	112.0	105.0
6	56.9	63.8	71.7	63.8
7	93.5	91.5	101.0	101.0

Table 2 *Reported concentrations of Tetryl (µg/g)*

	Low spike 1	Low spike 2	High spike 1	High spike 2
	20.1	20.1	25.2	25.2
Lab ID	**Concentration found by using the method**			
1	17.9	15.0	13.4	8.3
2	19.2	20.0	24.1	23.7
3	16.8	11.3	7.3	12.8
4	18.6	18.8	23.3	24.2
5	4.2	4.1	2.7	3.4
6	14.4	13.6	11.3	18.5
7	17.8	16.7	22.7	22.1

The graphs at the end of this activity—one for each analyte—were created from the data in the tables. You may never have seen information plotted in this way before, so it may take you a while to make sense of the graphs. The measured result for High Spike 1 is plotted against Low Spike 1, and High 2 vs. Low 2. This type of data display is called a *Youden two-sample plot* (one sample result plotted against a second sample result).

Key Questions

1. Look at the RDX graph. Each lab's results are indicated by a label on the points. How many labs contributed points to the graph? Why are there two points for each lab?

2. Instead of plotting in concentration units, the results are plotted as a ratio to the actual known concentration. Use the data in the table to calculate and verify the coordinates of the two Lab 6 data points on the RDX graph.

3. Explain why the numerical values for the axes on the graph have no units.

4. Describe what it means for a data point to fall exactly on the value (1.0, 1.0).

5. Each member of your group should take a turn to restate in different ways what the location of the Lab 6 points tells you about Lab 6. Do this in terms of the position of the two points relative to each other and to the (1.0,1.0) axis point.

6. Agree on and write down a conclusion regarding Lab 6's performance by using the two terms "random" and "systematic" to describe the performance.

7. Look for the Lab 5 data. Describe what you see. Discuss the location of the points in terms of the two terms "random" and "systematic". Agree on and write down a conclusion regarding the performance of lab 5 using the terms "random" and "systematic."

8. Reconsider the meaning of the Lab 6 and Lab 5 data points on the RDX graph and describe their position in terms of the words "percent recovery."

9. Look at the data for all other labs. Describe how each of these labs compares with Labs 5 and 6 in terms of random and systematic error and percent recovery. Note that there is an expanded version of the RDX graph which may help in answering this question.

10. Look up the terms random and systematic in your textbook to determine whether the way you have been using the terms is consistent with formal definitions.

11. What aspect of the location of points tells you about the accuracy of the work of an individual lab? What aspect of the location of points tells you about the precision of the work of an individual lab?

12. Discuss and write a written summary as a group of how a Youden two-sample plot shows evidence about the performance *of individual laboratories.*

Consider this...

The same labs also measured the quantity of tetryl in the same samples using the same method. Remind yourself of the structures of the two analytes. See the two tetryl graphs (full and expanded).

Key Questions

13. Describe the performance of each *individual lab* in terms of its accuracy and precision for measuring tetryl. (Each group member should take responsibility for doing this for two of the laboratories. Then, each group member should read his/her descriptions to the other group members and seek agreement about the description. The group should make a list and turn that in.)

14. Describe the performance of each individual lab in terms of random and systematic error and percent recovery. (Follow same procedure)

15. Write down a list of the procedural steps in the analytical method. If any steps are unfamiliar, refer to your textbook or consult your instructor.

16. Look at and compare the Lab 6 percent recoveries for RDX and for tetryl on the graphs. For both RDX and tetryl, the recoveries are in the range of about 60-70%. This suggests that a *common source of systematic error is affecting every measurement* for both analytes.

Each person in the group should individually try to identify and justify one specific reason—*that lies with the lab*—that could cause these observed recoveries of BOTH analytes. There must be a causal link between your reason and the location of the points on the graph. Then share and discuss your justifications and write down a list.

17. Identify and justify one possible specific reason—that lies with the lab—for why the Lab 5 results for RDX and tetryl are where they are on the graphs.

Consider this...

Consider the possible locations of points on a Youden plot relative to the axes.

A point in Quadrant 1 (upper right) is there because the lab had recoveries that were too high on both samples. A point in Quadrant 3 (lower left) is there because the lab had recoveries that were too low on both samples. Thus, points lying on this diagonal provide evidence regarding the level of systematic error.

A point in Quadrant 2 (upper left) is there because the first result was too low and the second too high. A point in Quadrant 4 (lower right) is there because the first result was too high and the second too low. Thus points lying away from the diagonal provide evidence regarding the level of random error. Note that random error is equally likely to cause results that are both low, or both high.

Turn your attention to the performance of the analytical method, start to finish. Remind yourself of the structures of the two analytes and the steps in the method.

Key Questions

18. Describe the performance of *the method* as a whole in terms of its success in providing reliable results for tetryl. You will need to compare the results *for tetryl* with the results for RDX. Agree on and write down your description, and the supporting evidence.

19. Identify and justify two possible specific reasons—that lie with the method—for the poorer performance on tetryl relative to RDX.

20. Discuss and write a summary as a group of how a Youden two-sample plot shows evidence about the performance *of an analytical method.*

21. As a group, summarize your perceptions. Were all group members improving in their ability to make sense of the graphical data?

Application

22. Assume that 10 labs were involved in an interlaboratory study of a new method. Sketch Youden two-sample plots for each of these cases. Each group member should take one case and then describe his/her graph to the others.

a. small random error and a wide range of systematic errors for labs

b. large random error and small systematic error for every lab

c. small random error and modest systematic error for every lab

d. small random error and large systematic error that affects every lab in a similar way.

23. A Youden two-sample plot requires only two samples per lab. We used two pairs of two samples. What information is retained and what information is lost by analysis of only two samples per lab?

24. As a group, outline and write down the steps needed to carry out an interlaboratory study that leads to constructing a Youden two-sample plot.

Graph 1 *Youden two-sample plot for RDX*

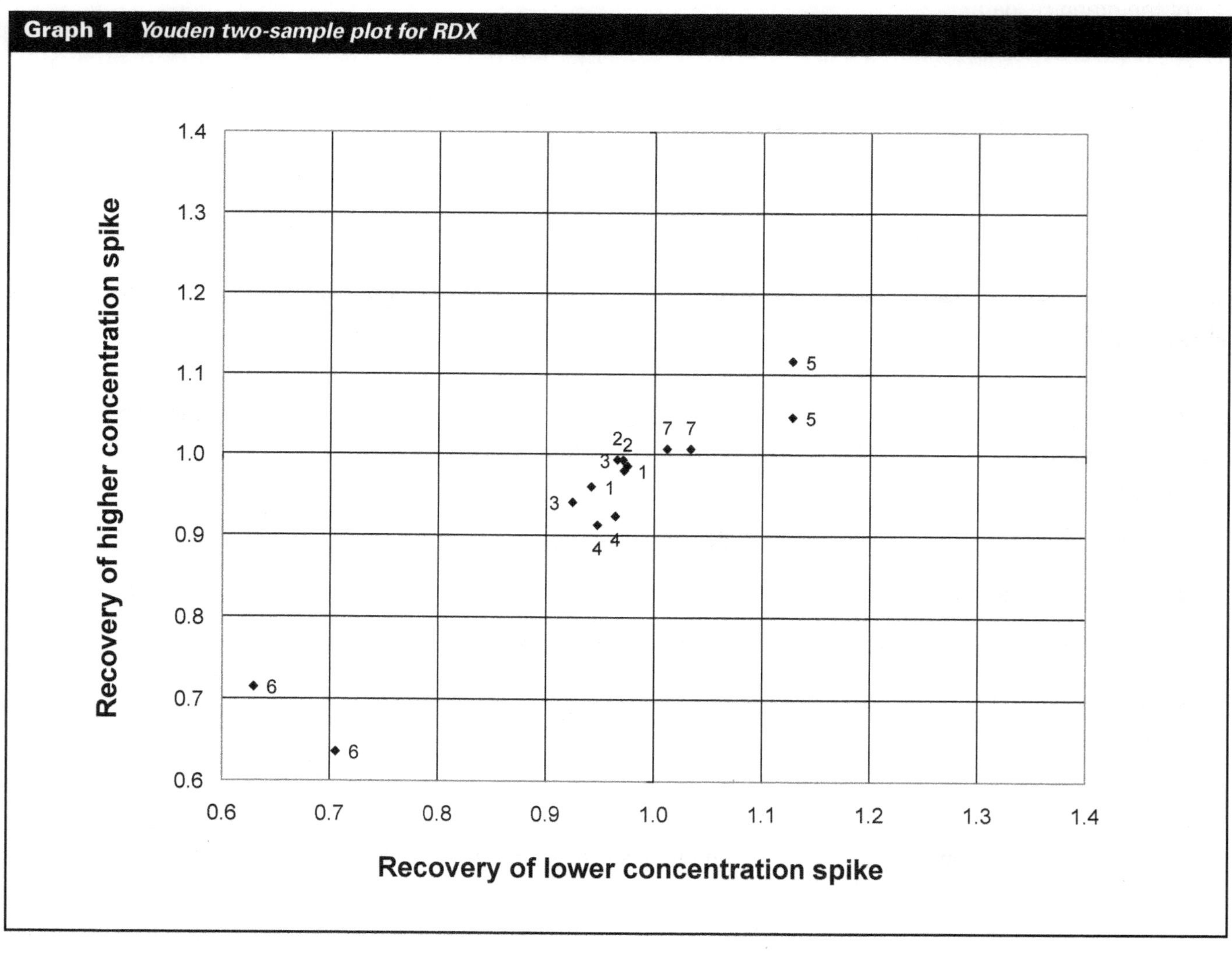

Graph 2 *Youden two-sample plot for RDX (expanded)*

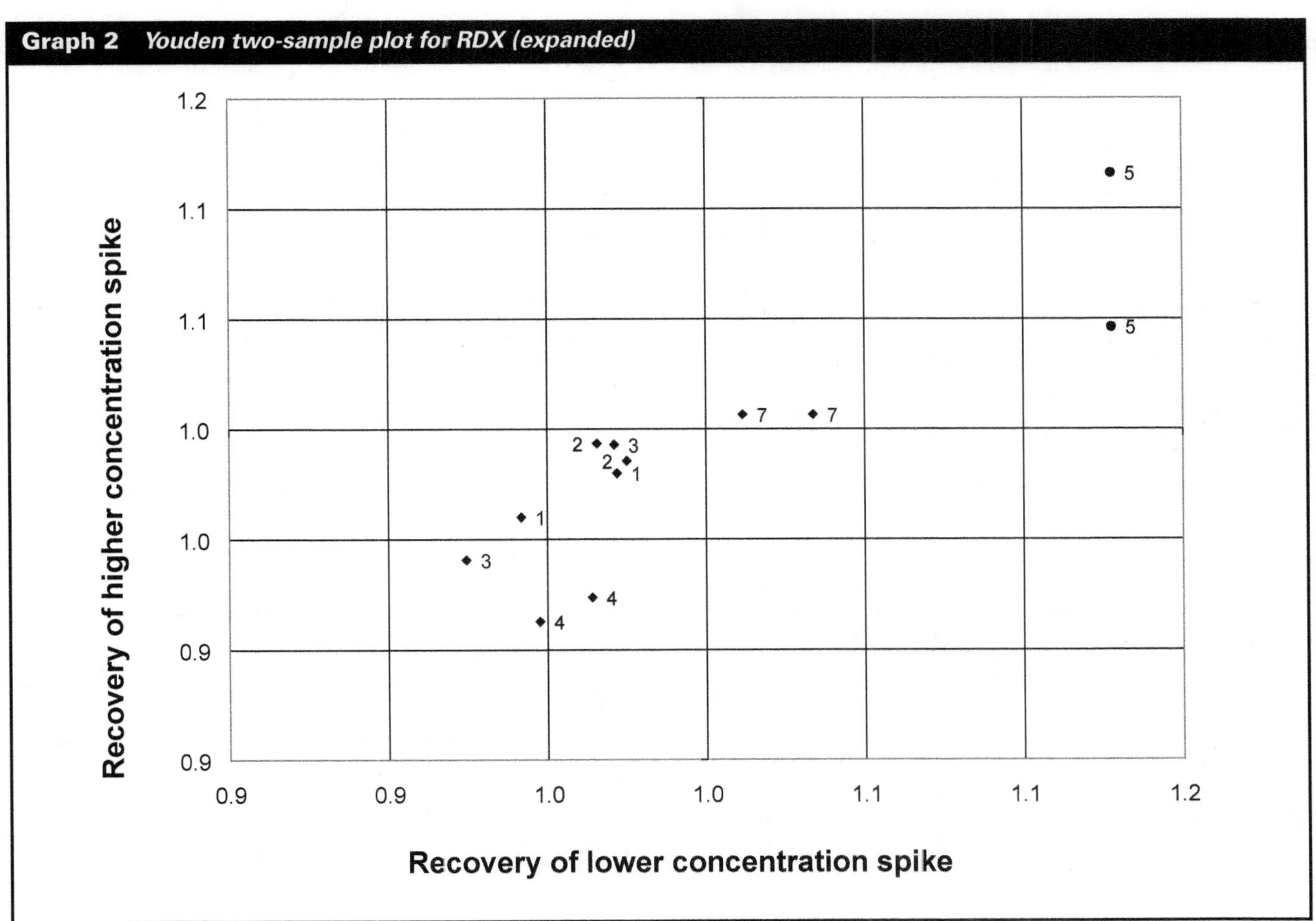

Graph 3 *Youden two-sample plot for Tetryl*

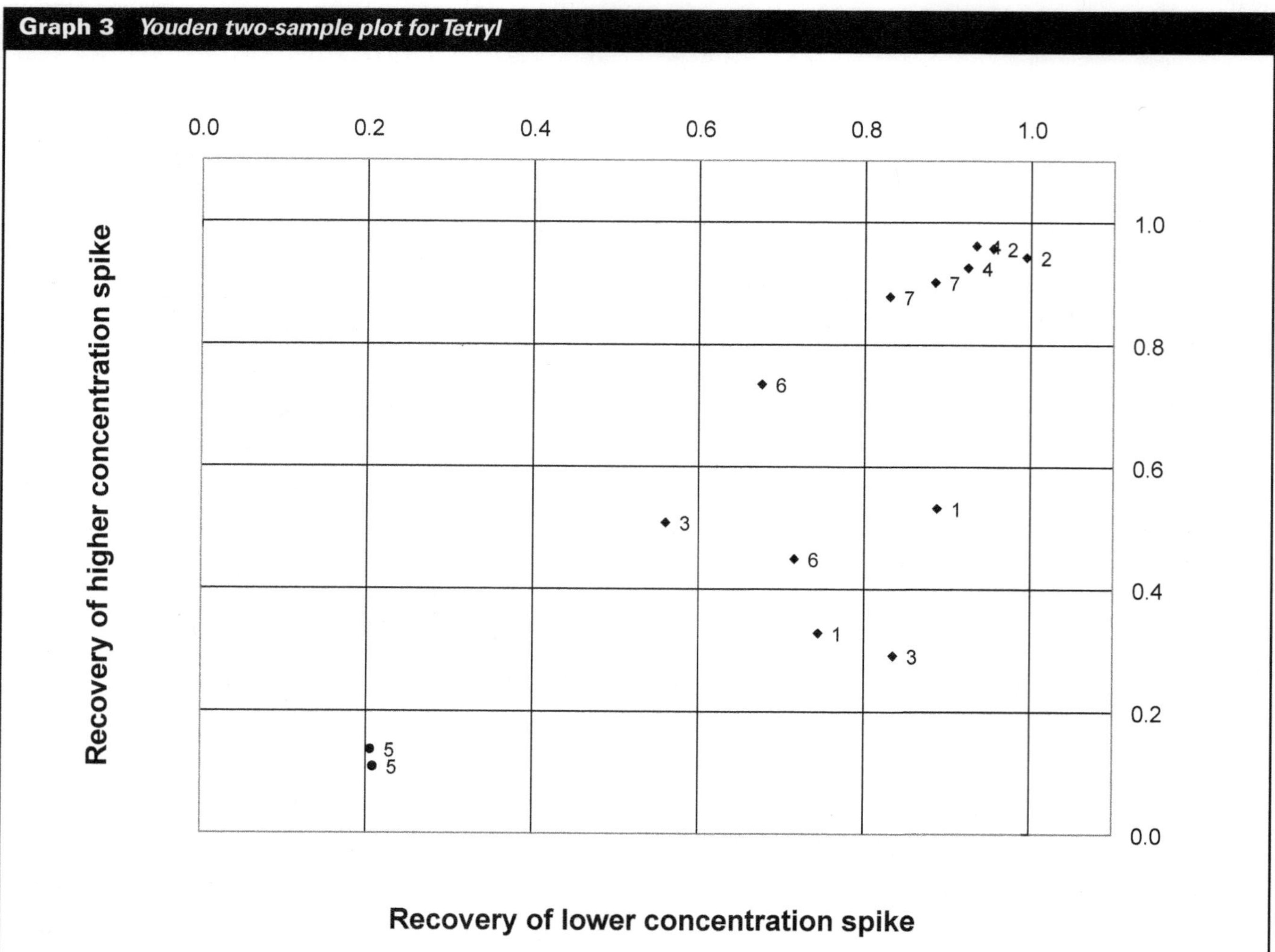

Graph 4 *Youden two-sample plot for Tetryl (expanded)*

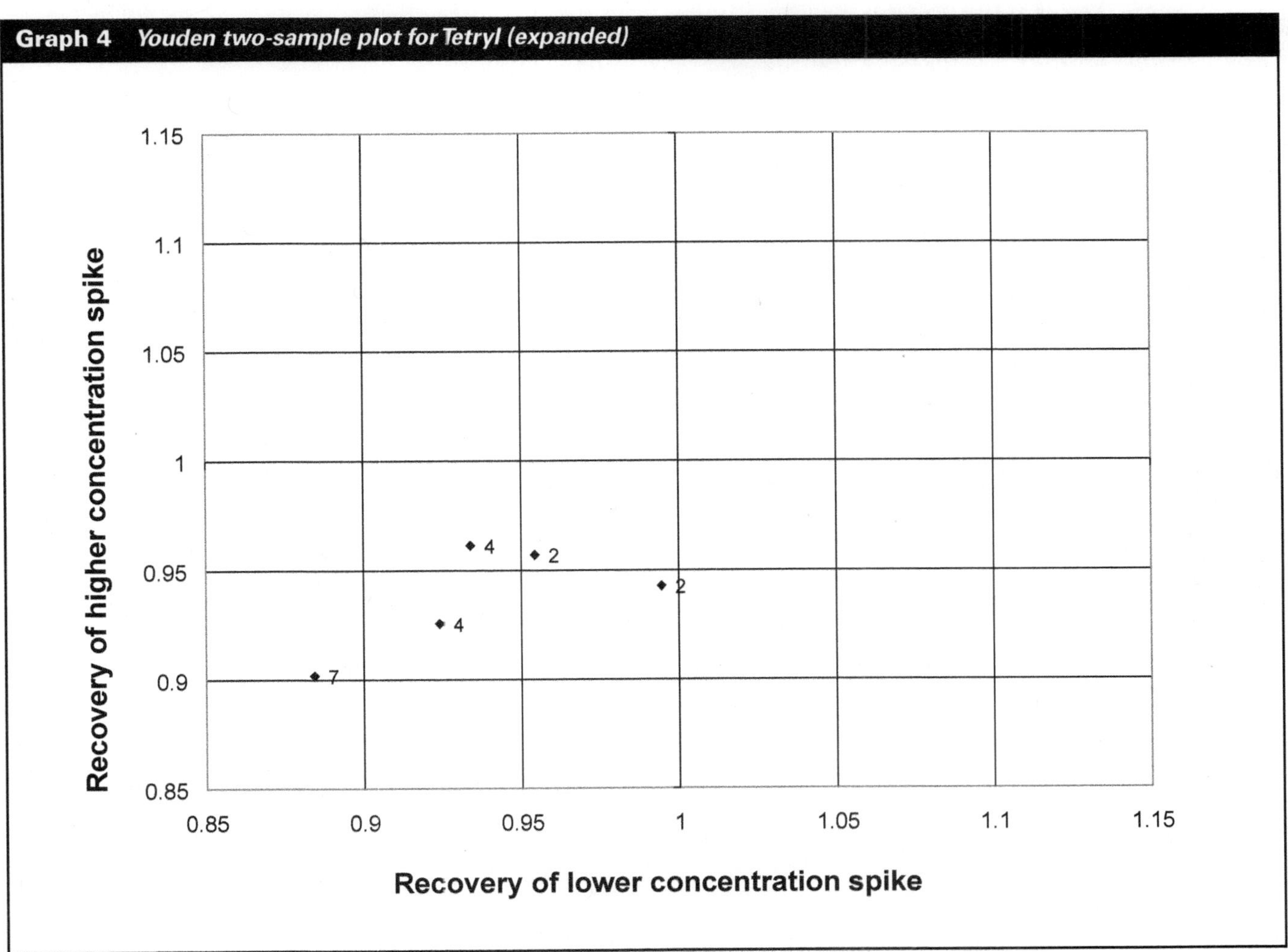

Instrumental Calibration: Method of Internal Standards

An Analysis Tool Used to Control for Instrument Variability

Learning Objectives

Students should be able to:

Content

- Describe calibration by internal standards as a method to deal with irreproducible instrumental results
- Calculate the concentration of an analyte in a sample when analyzed using the method of internal standards.
- Determine the need for the internal standard method.

Process

- Construct and interpret graphs. (Information processing)

Prior knowledge

- Instrumental calibration.
- Basic spreadsheet skills.
- Some basic knowledge of chromatography.

Further Reading

Harris, D. C. 2007. "Quality Assurance and Calibration Method." In *Quantitative Chemical Analysis,* 7th Edition, pp.90-91. New York: W.H. Freeman.

Author

Anne Falke

Consider this...

Acetylsalicylic acid (ASA) standards are prepared according to Table 1, each with 0.3 mg/mL salicylic acid (SA) added. They are analyzed by High Performance Liquid Chromatography (HPLC) using a C-18 column, and a mobile phase that is 70% 1M acetic acid/30% methanol. The results are found in Table 1.

Table 1 ***Results of HPLC analysis of acetylsalicylic acid (ASA) standards with 0.3 mg/mL salicylic acid (SA) added to each***

Run number	ASA concentration (mg/mL)	ASA peak area ($x10^{-5}$)	SA peak area ($x10^{-5}$)
1	1.00	6.329	1.918
2	2.00	8.061	1.257
3	3.00	9.237	0.975
4	4.00	23.062	1.968
5	5.00	16.284	1.053

The ASA calibration curve is plotted in Figure 1.

Figure 1 ***Plot of the ASA concentration vs. the instrumental peak area***

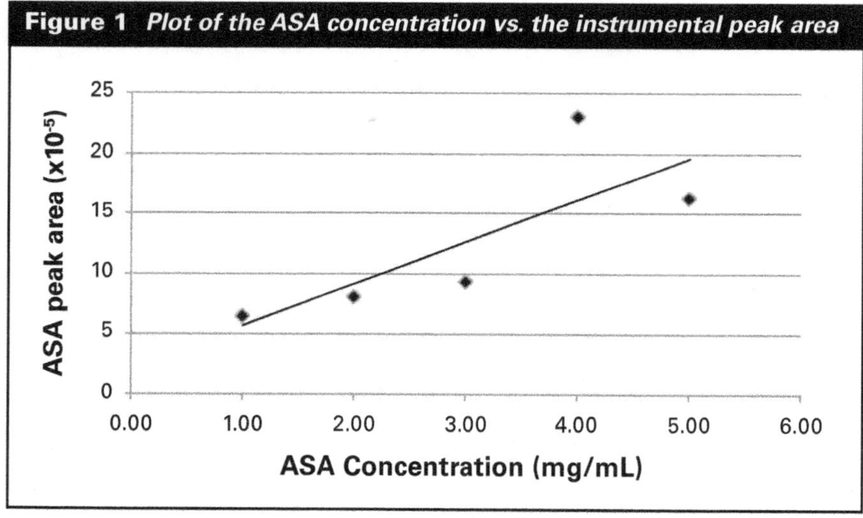

Key Questions

1. Describe the relationship of the data points to the trendline. Does the trendline adequately fit the data?

2. Should it be used to calculate the concentration of ASA in the samples? Why or why not?

Consider this…

Remember that a trendline is really an average. The distance from the trendline to the data point in the y-direction (vertical) is called the residual. The residuals indicate how far from the average or trendline each data point is. If a data point is on the trendline, its residual would be zero. If it is above the line, the residual is positive. Conversely, if the data point is below the line, the residual is negative. The residuals for the ASA calibration curve are plotted in Figure 2.

Figure 2 ***Residual plot for ASA calibration curve***

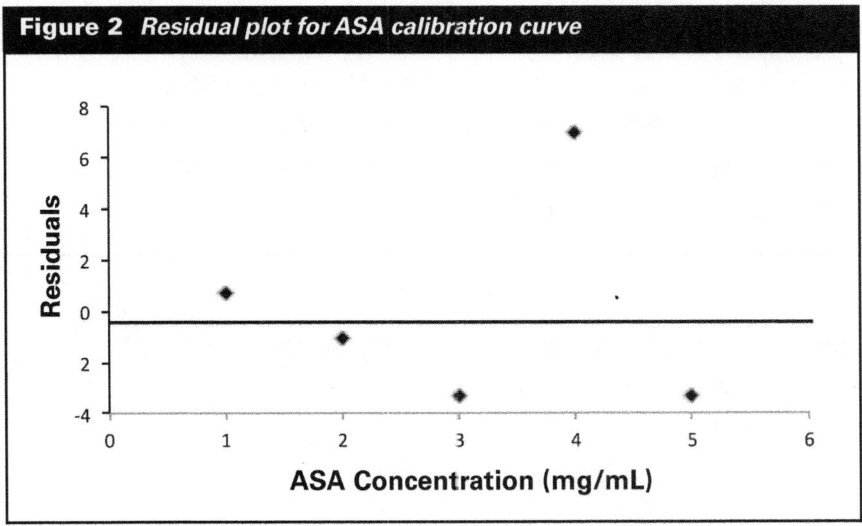

Key Questions

3. Examine the residual plot. What is the concentration of the data point at which the residual is closest to zero? Now look at that data point in the calibration curve. Is it closest to the trendline?

4. Which data point has the largest magnitude residual? Does that match the cata point furthest from the trendline?

5. The same amount of SA was added to every standard. How would you expect the peak area to compare from run to run? Do the peak areas match your expectation? Explain.

6. Calculate the average SA peak area. Calculate the difference between the actual peak area and the average. Plot the difference vs. the run number. This plot is also considered a residual plot. Compare the residual plot for the SA average to that of the ASA calibration. How does the residual pattern for the internal standard compare to the residual pattern for the ASA standards (Figure 2)? What does this imply?

The salicylic acid is used as an internal standard in the analysis of ASA. Both the ASA and the SA are analyzed in the same run, by the same instrument, so ideally, whatever affects the amount of sample reaching the detector will similarly affect both the analyte (ASA) and the internal standard (SA). Thus, one expects that both deviate from the average in the same way.

7. Now divide the ASA peak area by the corresponding SA peak area. Plot the ASA/SA ratio vs. the ASA concentration. Before continuing, compare your curve to that of another group.

8. What do you notice about this calibration curve? Should it be used to calculate the concentration of ASA in samples? Why or why not?

Consider this...

Two aspirin samples were prepared by grinding the aspirin tablet with a mortar and pestle, quantitatively transferring the powder to a 100-mL volumetric flask, dissolving in methanol, adding 0.3 mg/mL SA, and diluting to the mark with methanol. They were analyzed by HPLC using the same parameters as for the calibration standards. The results of the analyses are found in Table 2.

Table 2 ***Results of HPLC analysis of aspirin samples with 0.3 mg/ mL SA added to each.***

Sample	ASA peak area (x 10^{-5})	SA peak area (x 10^{-5})
Bayer®	13.16	1.874
Anacin®	13.40	1.861

Key Questions

9. Consider the calibration curve plotted in Q7.

a. What did you plot?

b. Given what is plotted in that calibration curve, decide as a group how to use the ASA and SA peak areas to obtain the concentration of the ASA in the samples.

c. Check your understanding. The concentration of ASA in the Bayer® sample is 2.24 mg/mL. Using what you decided upon in Q9b, calculate the concentration of ASA in the Bayer® sample. Did you get the same value? Now calculate the concentration of ASA in the Anacin® sample.

10. The aspirin tablets were each dissolved and diluted to 100 mL. What is the mass of ASA in each of the tablets? The manufacturer claims 225 mg ASA per tablet. Comment on the accuracy of the method.

The assumptions inherent in a standard calibration curve are:

- That the relationship between the concentration and the response is linear.
- That the response is reproducible (look at the salicylic peak areas and your answer to Q5).
- That there are no matrix effects.

11. Look back at the data in Table 1 and the calibration curve plotted in Figure 1. Which of these assumptions is/are not met? (NOTE: we haven't tested for matrix effects in this experiment, so for now assume there are none.)

12. Now look at the internal standard corrected curve. Which assumptions are not met by this curve?

13. Write a grammatically correct sentence to summarize when you would use the internal standard method of calibration.

Consider this...

Now that you know how to handle the math of internal standards, let's look at the selection of an internal standard. Below are structures for ASA and SA:

Figure 3 ***Chemical structures of acetylsalicylic acid and salicylic acid***

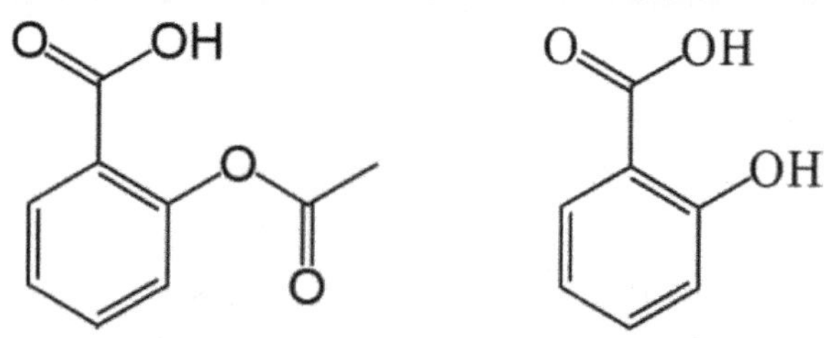

Key Questions

14. What do you notice about the two structures? Summarize the similarities and differences.

15. Why is it advantageous to use a compound with a structure similar to the analyte for the internal standard? Look at the information provided after Q6 if you need a hint.

16. If salicylic acid was already present in aspirin tablets in unpredictable amounts, how would this affect the response for the salicylic acid? What does this tell you about the selection of the internal standard?

17. As a group, summarize the criteria for selection of an internal standard.

Application

18. In addition to acetylsalicylic acid, caffeine is also a common component of many analgesics. In the solutions prepared to get the data presented in Table 1 the analyst also included caffeine standards. If the results for the analysis of standards and samples are as follows, what is the concentration of caffeine in the Bayer® and Anacin® tablets?

Run number	ASA concentration (mg/mL)	ASA peak area (x10^{-5})	Caffeine mass (mg/mL)	Caffeine peak area (x10^{-5})	SA peak area (x10^{-5})
1	1.00	6.329	0.100	4.779	1.918
2	2.00	8.061	0.200	6.204	1.257
3	3.00	9.237	0.300	7.509	0.975
4	4.00	23.062	0.400	19.06	1.968
5	5.00	16.284	0.500	13.16	1.053
6	Bayer®	13.16		11.659	1.874
7	Anacin®	13.40		14.142	1.861

19. In the analysis of samples for sodium by atomic emission, lithium is often used as an internal standard. Provide two reasons to use lithium as the internal standard. In the following analysis of sodium urine, what is the concentration of sodium in the sample?

Solution	Na emission	Li emission
5.0 ppm Na, 500 ppm Li	5.13	48.2
10.0 ppm Na, 500 ppm Li	7.62	36.4
25.0 ppm Na, 500 ppm Li	41.7	79.8
50.0 ppm Na, 500 ppm Li	62.5	57.7
Sample + 500 ppm Li	38.2	48.2

20. Gas chromatography(GC) typically uses the injection port in the split mode in which about 10% of the sample reaches the detector. Why is the use of the internal standard method particularly important with GC?

Sampling Error and Sampling Designs

Learning Objectives

Students should be able to:

Content

- Describe a sample as a part taken from the whole system of interest, where the part is representative of the chemical composition of the whole.
- Describe the different types of sampling designs and how they address compositional or temporal heterogeneity in the system of interest.
- Estimate the relative importance of sampling error relative to measurement error.

Process

- Seek patterns in tabulated data and summarize trends. (Information processing)
- Design a sampling strategy for a practical case. (Problem solving)

Prior knowledge

- Random and systematic error
- Mean, standard deviation, variance

Further Reading

- Jenkins, T.F., C.L. Grant, M.E. Walsh, P.G. Thorne, S. Thiboutot, G. Ampleman, and T.A. Ranney. 1999. "Coping with Spatial Heterogeneity Effects on Sampling and Analysis at an HMX-contaminated Antitank Firing Range." Field Analytical Chemistry and Technology 3 (1): pp. 19-29.
- Jenkins, T., T.A. Ranney, A.D. Hewitt, M.E. Walsh, and K.L. Bjella. 2004. "Representative Sampling for Energetic Compounds at an Antitank Firing Range." Cold Regions Research and Engineering Laboratory, US Army Corps of Engineers, ERDC/CRREL Technical Report, TR-047.
- Kratochvil, B., and J.K. Taylor. 1981. "Sampling for Chemical Analysis." Anal. Chem. 53: pp. 924A - 938A.
- Williams, B. *A Sampler on Sampling*. 1978. New York: John Wiley & Sons.

Author

Christopher F. Bauer

Consider this...

Political and consumer polls try to gauge public sentiment regarding likely voting or purchasing behaviors. News reports use the words "In a sample of 1000 likely voters who were asked whether they strongly agree, somewhat agree, somewhat disagree, or strongly disagree, 65% believe that (fill in whatever was asked) with a margin of error of 4%."

Assume the poll is intended to suggest what the entire voting population in the U.S. believes. The polling company will be concerned with the following potential sources of error:

a. The 1000 people selected do not provide answers that parallel the entire voting population.

b. The questions asked are misinterpreted by some of the people selected.

c. Individual people may say "strongly agree" at one time, but might have said "agree" at another time.

Key Questions

1. For each of these three sources of error, describe in your own words why each could be an error.

2. Extend your discussion in Q1 using the terms "Accuracy" and "Precision." Which sources contribute predominantly to inaccuracy in the conclusion? Which sources contribute predominantly to imprecision in the conclusion?

3. If you are the polling company, describe how much control you have over each of the sources of error. Discuss what you might do to minimize these sources of error.

4. All three of these can be called "measurement error," but because the sources are different, more descriptive terms can be used. Complete the middle column of Table 1 to show which error is associated with which term.

Table 1 *Measurement Error*

Error type	Situation (a or b or c)	Chlorophyll Measurement
Sampling		
Calibration		
Repeatability		

Any chemical analysis is intended to answer a question in order to make a decision. Potential sources of variation include the inherent variability of the instrumental measurement (random error or imprecision), the variability of the calibration due to unknown effects such as interferences (which often causes inaccuracy), and the variation in the composition of the samples removed from the population of interest. The last one is called "sampling error" – the extent to which the measured result obtained for the samples differs from the actual composition of the whole system of interest. Sampling error thus can cause the conclusion to be inaccurate.

5. Let's apply these ideas to a chemistry situation. Assume you want to add to our understanding of how the atmospheric concentration of carbon dioxide (a greenhouse gas) is affected by plant growth (which consumes carbon dioxide through photosynthesis). In particular, say you wanted to measure the concentration of chlorophyll in plant leaves in the Amazon River Valley to determine the total average chlorophyll concentration.

   There are three potential sources of error (x, y, z) listed below. Each group member should consider privately where these errors fit in the table above. Then compare your choices. Discuss which matches make sense; then come to an agreement and complete the table.

   x. The instrument that measures chlorophyll unknowingly also responds to xanthene (another common component in plant leaves).

   y. One leaf from each different type of plant is obtained, measured for chlorophyll, and averaged without regard to the location or abundance of each plant type.

   z. When chlorophyll concentration is measured with an instrument, the result is subject to instrumental noise variation in response.

An analyst must make a good decision about which portions of the system to select for measurement (because decisions will be based on the analytical results). One major goal in collecting samples, then, is to make sure that the samples are *representative* of the system as a whole. This can be accomplished by selecting an appropriate sampling strategy.

Consider this…

A number of sampling strategies may be implemented.

Strategy	How implemented
Grab	Sampling sites selected for convenience in location or time
Random	Sampling sites identified by random selection from all possible sites
Systematic	Sampling sites identified at regular intervals of location or time. Also called "periodic."

We will use a farming scenario to explore how each strategy works and the relative advantages and disadvantages.

Your company has been hired to establish the best strategy for measuring pesticide residues in fields across an entire county. A pesticide (N-methyl carbamate) was used during growing season. You have selected one field site (1 acre area) for testing. You have divided the field using wooden stakes into a 10x10 grid of equal-sized squares (each about 6 m across). (See Figure 2 on next page). Each square can be identified by a column and row. For example, K10 is in the lower right corner. One soil sample will be removed with a soil auger (Figure 1) from the center of a grid square.

The group manager should request a grid showing the actual concentration of pesticide residue in each grid location (in units of nanograms per gram soil). The manager should NOT SHOW this data to the rest of the group yet. This information is in the Instructor's Manual.

***Figure 1** US Dept of Agriculture photo of several different soil augers.*

(Do a web search for "soil auger" images to see many examples.)

Courtesy of Natural Resource Conservation Service, United States Department of Agriculture

Figure 2 *Overhead map of staked field divided into 10 x 10 grid (6 meter squares)*

A											
B											
C											
D											
E											
F											**Truck road between fields**
G											
H											
J											
K											
	1	2	3	4	5	6	7	8	9	10	

6. Consider the following strategies. As a group, come to an agreement on which grid locations could possibly be sampled for each strategy. In particular, discuss how strategies A and B will be accomplished other than by human intuition (which has difficulty creating random patterns). [Your instructor's assistance may be needed.]

   a. 10 random samples

   b. 5 random samples, each of which is split in half and analyzed separately

   c. 10 samples on horizontal transect*

   d. 10 samples on a vertical transect*

   e. 10 samples on a diagonal transect*

   f. 5 samples equally spaced on a vertical transect*

   g. 3 samples along the road

   h. 1 sample from center of field

   *A *transect* requires moving in a direct line from one side to the other across the target area.

7. Assign each group member one or more of the strategies. Each person should identify grid locations to sample (using the blank grid), and submit those to the manager (who will send back the analytical results written on the grid). Each person should then calculate privately the mean and standard deviation of their sample results.

8. Enter your data into the appropriate row in Table 2. Complete the table with the results from your group.

Strategy	**Type**	**Grid positions sampled**	**Number of samples (n)**	**Mean concentration of N-methyl carbamate (ng/g)**	**Standard deviation (ng/g)**			
A								
B								
C								
D								
E								
F								
G								
H								
Complete								

9. Add to the table the type of strategy: grab, random, or systematic.

10. Which strategy or strategies (A through H) can provide direct information about instrumentation measurement error separate from sampling error? Explain.

11. Calculate the pooled standard deviation for the duplicate results for strategy B. (Shortcut: calculate the sum of the differences squared divided by 2.) This is an estimate of the instrumentation measurement error. Compare this number with any of the standard deviations listed in Table 2. Based on what you see, how large a contributor to the overall variation in results is instrumental measurement error? Explain.

Often, instrumentation measurement error is smaller than the variability across samples. Consequently, it is more useful to take more samples than to replicate measurements on the same sample.

12. Looking only at the results in this table and not considering the individual grid concentrations, compare the mean values and standard deviations. Each group member should take a turn proposing at least one observation, or question, or speculative explanation, about what you see for the different sampling strategies. Write down at least as many things as there are group members.

13. Are you able, at this point, to suggest one or more strategies as best? If so, which and why? This is not a final decision, just an initial judgment.

14. Are you able to suggest one or more strategies as inadequate? If so, which and why?

15. The manager should obtain from the instructor the mean and standard deviation result for the analysis of all 100 grid locations. Add this information to the table on the line labeled "complete." Which strategy or strategies now look "best"? Explain in terms of the concept of *representativeness*.

16. Think about actually being in the field, organizing the sampling activity, moving around the field, and collecting samples. For the strategies you contributed to Table 2, estimate the number of minutes it would take to accomplish each one, if you had to go to that field and implement it. Explain your estimates to the group. Obviously, you can't know exactly how long it might take until you do it, so just try to agree on reasonable time estimates. Add this information to one of columns on the right side of the table.

Figure 3 ***Corn field after harvest***

Courtesy of Natural Resource Conservation Service, United States Department of Agriculture

17. Assume that the cost of a single analysis of a soil sample is $100. Assume the cost of your time is $50 per hour (which includes your salary, transportation, overhead, etc.). Assume there are 100 fields in the county that need to be studied. For the strategies you contributed to Table 2, calculate the analytical cost and the sampling cost. Enter the two pieces of data into the last two columns in the table.

18. Recall that the ultimate goal of the project is to use this strategy in many locations across the county. Based on the additional information in the last columns, which strategy or strategies seem best? Explain.

Systematic Sampling

19. Taking fewer samples will reduce cost. Compare the costs for systematic strategies D vs F (which takes half the samples). Describe the differences in the analytical results. The differences demonstrate a problem that may arise in systematic strategies. What do you think is the problem? It may help to look at the overall mean and standard deviation for all 100 sites.

20. Think like a farmer. Think about how the pesticide was probably applied. The manager should reveal the entire grid of results at this time. Look at the pattern of pesticide residue concentrations across the entire field. How might that have come about?

21. Review your suggestion in Q19 in light of what you've now seen in Q20.

Systematic sampling approaches are subject to being biased when the analyte of interest varies at a regular frequency that is the same as the sampling frequency. Sampling at more than twice the likely frequency of compositional variation will account for this type of problem (assuring that peaks and valleys are seen). Consequently, in a totally unfamiliar situation, it is safer to sample more frequently at first, and then to reduce the rate once the system is better understood.

Grab (Convenience)

22. Look at Table 2. Comment on the advantages and disadvantages of using a grab sample strategy.

23. Review the result for a single sample taken midfield (Strategy H). It's not very reliable. However, in the right circumstance, a single sample could suffice. What would that circumstance be?

24. Review the results for the three samples taken along the road (Strategy G). The average is not a good representation of the whole. However, in the right circumstance, this strategy could suffice. What would that circumstance be?

25. Draw a conclusion about when it would be appropriate to rely on grab samples. Explain.

The challenge is to determine whether the system of interest is *homogeneous* (in which case a single sample would be representative) or *heterogeneous* (the composition is different in different locations or at different times, thus requiring multiple samples). Knowing something about the domain from which samples must be collected (e.g., cornfield) will aid in making appropriate choice of strategy.

When substantial heterogeneity is certain or highly likely within different zones at the sampling site, then those zones can be sampled and analyzed independently. This strategy is called *stratified sampling* (from "strata" or layers, as in sedimentary rock deposits.) Within each zone, any one of the previously discussed strategies (grab, random, systematic) can be used. Sometimes results for each zone are reported separately. Sometimes results from each zone are averaged but the average must be weighted according to some physical characteristic, like relative area.

26. Given what you now know about the distribution of pesticide residues in this test field, describe a procedure that takes a stratified approach. Determine the number of samples to take, from which locations, and how to combine the data to represent the whole field. Add your result to Table 2.

Composite Samples

So far, we've made the assumption that we would take one sample from each sampling location and perform a single chemical analysis on that sample. An alternative strategy is to collect several samples, but instead of analyzing each, combine them, mixing well, and remove one portion of that mixture for analysis. This is called a *composite sample*. Usually, only a portion of the composite sample is analyzed.

27. Discuss what you gain and what you lose with this approach.

28. Which of these situations would be amenable to creating a composite sample? In each case, describe at least one qualifying condition. (i.e. complete the sentence ". as long as ..."). Each group member should take one of these situations and lead the discussion.

- Measuring average pesticide residue in a corn field.

- Measuring total calcium transport in a river entering an estuary.

- Measuring lead in blood taken from children in a particular school.

- Measuring average carbon monoxide exposure of urban residents.

This suggests that you are not interested in the variation among individual samples, but an average. This also requires that the mixing process results in a homogeneous mixture. This is easily achieved with gas phase and liquid phase materials, but can be problematic with solid materials unless they are reduced to small particle size, and mixed and split in an unbiased manner. Techniques such as cone-and-quarter, and devices such as a riffle (which splits a powdered sample into two equivalent parts) are used to achieve this latter goal.

Summary

29. Go around the group and ask each group member to mention one thing he/she learned about sampling. Keep going around until no one has anything more to add. Write a summary list here.

30. Based on what you've learned, make a list of advantages, disadvantages, and cautions for each sampling strategy (grab, random, systematic, stratified, composite).

31. Comment on how well the members of your group achieved the process goals for this activity:

- How confident are individual group members in seeking patterns and summarizing trends in tabulated data?

- How confident are individual group members in designing sampling strategies?

Application

32. Use the Internet to find out what a "riffle" is and what the "cone and quarter" procedure is.

33. What type of sampling strategy is being used in each of the following cases? Explain.

a. Person with diabetes tests blood sugar every hour.

b. Police officer administers a breathalyzer test to a suspected drunk driver.

c. Single bottles of soft drink are removed at preset random intervals from a production line for quality control analysis.

d. An estimate of exposure of urban residents to diesel exhaust particles is desired.

34. In which of the cases in Q33 would you be most interested in separating the sampling error from the analysis error? Explain why, and how you would accomplish this.

35. You have 100 corn fields to study. Discuss how you might use grab sampling as a means for screening as a preliminary step to determining which fields to sample most intensively.

36. If the distribution pattern that you found for this test corn field (Figure 2) was known to be reliably true for all fields, it would be possible to take samples from just three sites in the 10x10 grid and calculate a weighted average pesticide residue for that field. Discuss with your group and speculate as to how this would work. Use your ideas to perform a calculation, and verify that the result is valid by comparing with the exhaustive sampling mean. What type of strategy is this?

37. Explain why calculating a simple average of the three samples described in Q36 would not be a valid way to represent the composition of the entire field.

38. Earlier you were told that $\Sigma\ (d^2/2)$, where d is the difference between duplicates gives the pooled standard deviation. Verify this by using the formal approach: square root of sum of squares of the standard deviations from duplicates.

39. How could you modify Strategy F so that you protect yourself against the problem you found (when the frequency of sampling matched the frequency of variation of concentration of pesticide), but still retained the cost savings of collecting and analyzing fewer samples?

40. Because of your experience with pesticide residues in fields, you are asked by the local government to design a sampling protocol for monitoring the concentration of a different pesticide residue in a pond located in the same county as the fields you have been sampling. Based on what you know right now, list the five most important questions that you consider critical to answer or to get information about BEFORE you do any sample collection or analysis.

Errors in Measurements and Their Effect on Data Sets

Learning Objectives

Students should be able to:

Content

- Classify different sources of experimental error and determine their effect on the mean and dispersion of the data.
- Develop the ability to judge whether an analytical balance has been used correctly.
- Define situations in which it is reasonable to reject data.

Process

- Develop critical skills in assessing the quality of data: Are they "good" or "bad"?
- Identify qualitative differences among data sets.
- Develop generalizations from tabulated data.

Prior knowledge

- Familiarity with simple statistics terms such as mean, range, and standard deviation.
- Familiarity with an analytical balance that can be tared. In particular, students need to know that closing the draft shield improves the reproducibility of the balance, and that hot objects placed on the balance pan create convection currents that reduce the measured value of the mass. They should also know that analytical balances need to be leveled for correct operation.

Further Reading

- Harris, D.C. 2007. *Quantitative Chemical Analysis,* 7th Edition, 42. New York: W.H. Freeman.
- Taylor. John R. 1997. An Introduction to Error Analysis, 2nd Edition, 94. Sausalito CA: University Science Books.

Author

Carl Salter

Consider this...

As scientists we often must make a large number of repeated measurements. It is usually not practical to communicate or transmit every single measurement to other scientists who are interested in the experiment; instead, we summarize the data set using carefully selected statistics. There are two key aspects of the data: one is the central tendency, or center of the data, which is a value about which the data seem to "cluster," and the other is the spread or dispersion of the data. To summarize the center of the data we usually use the average (or mean) value. The average is the sum of all measurements divided by the number of measurements. To summarize the dispersion of the data we can use either the range or the standard deviation. The range is simply the maximum minus the minimum. The standard deviation is more complicated, and involves the sum of the squares of the differences between each measurement and the average.

Here are formulas for the average, the range, and the standard deviation.

$$\overline{x} = \frac{1}{n}\sum_{i=1}^{n} x_i, \quad r = x_{max} - x_{min}, \quad s = \left(\frac{1}{n-1}\sum_i (x_i - \overline{x})^2\right)^{1/2}$$

Since 1982, U.S. pennies have been minted from 97% zinc stock electroplated with a thin coat of copper; a typical new penny weighs 2.508 grams. Prior to 1982, pennies were made almost entirely of copper, and weighed more than 3 grams.

Suppose that for several months, a professor has collected lots of pennies from his loose change to use in an experiment that helps students learn to use analytical balances. He removes the pennies that are dated 1981 or earlier and keeps the more recent pennies for the experiment. The pennies are divided into sets of 20.

On the day of the experiment, a trained laboratory assistant takes one of the sets of 20 pennies and weighs them on an analytical balance with a digital display that weighs to 0.0001 grams. Each student then weighs a different set of 20 pennies and records the weights in a laboratory notebook. The students also use analytical balances that weigh to 0.0001 grams.

The students are asked to compare their data with the data obtained by the lab assistant.

Key Questions

1. The following table contains three sets of penny weights; the first set is the set of measurements made by the lab assistant. The other two sets were obtained by students. The data are listed in the order in which the measurements were made. In addition to the raw data, the table contains the summary statistics for each set: average, range, and standard deviation. Compare the three data sets.

Masses (grams) of pennies			
	Lab asst	**Set 1**	**Set 2**
	2.5283	2.4814	2.5696
	2.4936	2.4782	2.4200
	2.5442	2.4989	2.5234
	2.4892	2.4948	2.4650
	2.5099	2.5065	2.4597
	2.5058	2.4392	2.5110
	2.5580	2.4946	2.5789
	2.5534	2.4932	2.5519
	2.5035	2.5332	2.4505
	2.5058	2.5177	2.5205
	2.5264	2.5423	2.4875
	2.4936	2.4826	2.4896
	2.5038	2.4801	2.5029
	2.5238	2.5022	2.4534
	2.4911	2.4989	2.5559
	2.4892	2.4392	2.5096
	2.4924	2.5470	2.5013
	2.5058	2.5424	2.5673
	2.5580	2.4925	2.4850
	2.5534	2.4910	2.4945
Average	**2.5165**	**2.4978**	**2.5049**
Range	**0.0688**	**0.1078**	**0.1588**
Std deviation	**0.0248**	**0.0293**	**0.0438**
Explanation			

There are important differences between the lab assistant's data and the data obtained by the students. Which of the following explanations fits the differences among the sets? Write your group's choice on the bottom of the table under each of the three columns of data.

Warm object: A student keeps all the pennies in his pocket as he weighs them; as a result, the pennies are above room temperature when they lie on the pan and give low measurements.

Draft: One student doesn't completely close the draft shield as she weighs all 20 pennies.

Wear and tear: Different pennies will receive different levels of wear, dirt, oil, and corrosion while in circulation.

Using a complete sentence, write down a reason for your group's selection for each set.

Lab asst.:

Set 1:

Set 2:

Here are three more sets of data obtained by students who made some mistakes. Compare them to the set from the lab assistant, which is listed again for your convenience. Examine the sets and match them to an explanation.

Masses (grams) of pennies				
	Lab asst	**Set 3**	**Set 4**	**Set 5**
	2.5283	2.5132	2.4969	2.4236
	2.4936	2.5287	2.4864	2.5144
	2.5442	2.4944	2.5453	2.4879
	2.4892	2.5238	2.4812	2.4911
	2.5099	2.5175	2.5044	2.5058
	2.5058	2.4892	2.5005	2.5535
	2.5580	2.5099	2.4746	2.5533
	2.5534	2.5085	2.4864	2.5020
	2.5035	2.5132	2.4978	2.5058
	2.5058	2.5535	2.5455	2.5257
	2.5264	2.4502	2.4958	2.4924
	2.4936	2.5020	2.5207	2.4892
	2.5038	2.9464	2.4069	2.5124
	2.5238	2.5257	2.4955	2.5257
	2.4911	2.5175	2.5095	2.4936
	2.4892	2.5264	2.4812	2.4911
	2.4924	2.4879	2.5019	2.5132
	2.5058	2.4780	2.5005	2.5099
	2.5580	2.4936	2.4069	2.4502
	2.5534	2.5085	2.4969	2.5056
Average	**2.5165**	**2.5294**	**2.4917**	**2.5023**
Range	**0.0688**	**0.4962**	**0.1386**	**0.1299**
Std deviation	**0.0248**	**0.1006**	**0.0344**	**0.0295**
Explanation				

Not level: One of the balances is not level; as a result it gives measurements that are slightly low.

Not tared: One student fails to tare the balance when she weighs her first penny, but thereafter she tares the balance between each weighing.

Typo: A student transposes the second and third digits of one value as he records it in his notebook; that is, he records 2.94xx instead of 2.49xx.

Using a complete sentence, write down a reason for your group's selection for each set.

Set 3:

Set 4:

Set 5:

2. Which error had the greatest effect on the average value? How do you know? Why did it have a big effect?

3. Are the range and standard deviat on consistent? That is, for the six sets of data, do the range and standard deviation show similar trends regarding the dispersion of the data? Explain.

4. Which error had the greatest effect on the dispersion of the data? How do you know? Why did it have a big effect?

5. Compared to the lab assistant's data set, what happens to the statistics of the other sets of data? Your group should discuss each data set. In the table below, summarize the changes in the average and the standard deviation for each cause. Indicate whether the statistic was affected, and whether it increased or decreased. Use up or down arrows to indicate increase or decrease.

Explanation/cause	Average	Std Deviation
Warm pennies weighed		
Draft shield open		
Not level		
Not tared for first penny		
Typo: transposed digits for one value		

Consider this...

All measurements are subject to error. Different sources of error will affect the data set differently. An experimental error will usually have one of the following effects:

1) Affect just one measurement in a unique way, which to some degree can change both the average and the dispersion,

or

2) cause an increase in dispersion of the data by affecting all the data in unpredictable, random ways,

or

3) cause all the data to shift either to higher or lower values by about the same amount, which, of course, will have an effect on the average. This effect can be eliminated if the source of the error can be found.

Key Questions

Q6 - Q8 refer to the information about effects of error given above.

6. Will the second type of error affect the mean? If so, how will it affect the mean?

7. How will the third type of error affect the dispersion of the data as measured by the standard deviation?

8. If an instrument is used properly, what type of error will still be present in the data?

Now apply the information about effects of errors to Sets 1-5 of the penny data.

9. Assign causes to an effect.

Effect	Explanation/cause (from table above)
Affect average and dispersion by changing one datum	
Increase dispersion by altering all data	
Shift all values, changing the mean	

10. Which of these three types of error would you call a blunder? Which is a systematic error or bias? Which is random error?

11. In a complete sentence, summarize the effect of a blunder on the summary statistics of the data.

12. Summarize the effect of a systematic error on the summary statistics of the data.

13. Summarize the effect of random error on the summary statistics of the data.

14. Why is random error always present even when an instrument is used correctly?

15. Explain how you could spot a systematic error or a blunder. Which of these two types of error is easier to spot?

16. What types of error require a student to discard data and make a new set of measurements?

17. Many students will say that a data set has been affected by "human error," but most scientists do not accept "human error" as a type of experimental error. What do you think "human error" means, and what effect would it have on a data set?

As a group, discuss the three types of errors and decide which, if any, are "human error."

Application

18. a. On many rulers, the zero mark for the measurement scale is not at the physical edge of the ruler. What kind of error will result if someone uses the ruler and doesn't notice this? What will be the effect on the data?

b. When you use a ruler, you sometimes have to estimate the length between scale markings. If you do this several times, your estimate will probably not be the same every time. What kind of error is this? Explain its effect on the data.

19. Very often, scientists will check or calibrate an instrument to make sure it is making correct measurements on an accepted standard sample. What type of error does this practice reduce or eliminate?

20. Modern instruments usually have digital readouts; the reading in the rightmost digit will sometimes fluctuate up and down among several values. What type of error is this? Explain its effect on a large data set.

21. What type(s) of error(s) does the control experiment reveal? Explain.

22. Suppose that you can reproduce a short-time physical or chemical event over and over again. Discuss possible sources of errors associated with measuring the duration of the event using a stopwatch. ("Human error" is not a source.) What effect do the sources of error have on the data?

23. a. How do scientists measure random error? What statistical values express the amount of random error in a data set?

b. How do scientists measure systematic error? What statistical values express the amount of systematic error in a data set?

24. What are the advantages and disadvantages of using the range to estimate dispersion? Contrast it to the standard deviation as a measure of dispersion.

25. Suppose you are a teaching assistant for the penny experiment, and your job is to review the student's data for possible problems. What would you expect to see in a student's data set for the following problems? What course of action would you suggest to the student?

a. A student misreads the balance and puts the decimal point in the wrong place for all 20 measurements.

b. One student weighs all 20 pennies without checking to make sure the balance reads zero between each weighing.

c. One student misunderstands the instructions and weighs one penny 20 times.

d. A student doesn't want to wait for an analytical balance, so he finds a balance that weighs to 0.01 grams and uses that balance to weigh the pennies.

e. A student gets one penny that was minted early in 1982, when some pennies were still made with 95% copper.

26. "Your data are precise but inaccurate." What have you just been told about the random error and systematic error in your data?

27. "Your data are imprecise but accurate." What have you just been told about the random error and systematic error in your data?

28. Suppose you and two lab partners need to weigh six pennies quickly; you have access to three balances so each of you weighs two pennies. Suppose exactly one of the six pennies is an old one made of copper, but no one in your group knows about the change in penny composition that took place back in 1982. Would the weight of this penny look like a blunder, a systematic error, or a random error? How would you try to determine the source of the error? Ultimately, how would you convince yourself that one penny really weighs more than the others?

Linear Regression for Calibration of Instruments

Learning Objectives

Students should be able to:

Content

- Describe linear regression as an averaging process.
- Interpret the results of a linear regression analysis.
- Evaluate the quality of a linear model for calibration of an instrumental measurement.

Process

- Construct and interpret graphs (Information processing)

Prior knowledge

- Mean, standard deviation.
- Instrumental calibration.
- Basic spreadsheet skills.

Further Reading

- Standard Quantitative Analysis Textbooks
- Anscombe's Quartet (classical model of problems with correlation coefficients as measures of fit) http://en.wikipedia.org/wiki/Anscombe%27s_quartet (accessed 8-17-13).

Author

Christopher F. Bauer

In a previous activity, you learned how to calibrate an instrument and evaluate figures of merit such as dynamic range and detection limit. We are revisiting calibration in order to explore the statistical procedures involved.

We'll begin by revisiting another idea with which you are already familiar—the mean or average—but with a new interpretation that will be developed through the Key Questions below.

Consider this...

A laboratory technician has been given a sample of household tap water in which to determine the concentration of lead. She removes six subsamples and determines the lead concentration in each using furnace atomic absorption spectrometry. (This instrumental measurement is based on the amount of light absorbed by atoms of lead produced by the furnace. The furnace is raised rapidly to a temperature of more than a thousand degrees C to decompose and vaporize the lead in a microliter-sized volume of sample.) The instrument signals and summary statistics are in Table 1.

Table 1 ***Absorption measurements for lead in a single sample***

Trial	Signal in absorbance units
1	0.033
2	0.040
3	0.030
4	0.039
5	0.034
6	0.034

Mean	=	0.035
Standard Deviation	=	0.0038
Variance	=	1.44E-05
Sum of Squares	=	7.714E-05

Key Questions

1. Sketch a graph of signal magnitude vs trial number.

2. Draw a horizontal line on the graph at the value for the mean.
Infer from the graph the values for the slope and intercept for this line.

3. As a group, decide how to describe the graph and its points in words, as if telling another person who cannot see the graph. Don't just repeat numerical values. Describe the position of points and the line. Write this down. Exchange your description with another group to check for understanding.

4. Calculate the differences between each point (y_i) and the mean ($\overline{y}$, called y bar). Draw lines on the graph to represent these differences.

5. Square each of these differences and add them together. Write an algebraic representation for this process. This is called the "sum of squared deviations from the mean" or simply the "sum of squares," sometimes abbreviated as SS.

6. Describe the ways in which the graph appearance would be different if the data set had a larger sum of squares, but the same mean. What would happen if it had a smaller sum of squares but the same mean?

7. Assume four other analysts approach you, each with a different suggestion about the appropriate central value for the data set: 0.030, 0.032, 0.038, or 0.040 absorbance units, respectively. Have each member of the group calculate the sum of squares for the data based on one of the four proposed central values.

8. Construct a graph of sum of squares vs. proposed value of mean. Include in the graph the original mean and sum of squares from Table 1 and the values calculated in Q7.

9. In your own words, describe the shape of this graph. Try to recall the mathematical term for the shape you see.

10. Consider the shape of the graph. What is unique about the correct mean (0.035 absorbance units) as an indicator of the central tendency of a set of measurements? Agree on and write down one or two concise sentences that summarize this idea.

Consider this...

Returning to the problem of measuring lead in household tap water by atomic absorption (AA) spectrometry, the laboratory technician calibrated the AA instrument by measuring the absorption signal for a set of standards containing different lead concentrations. The data are shown here in Table 2 and in Figure 1.

Table 2 ***Absorption calibration data for lead***

Lead Concentration (μg/L)	10.0	20.0	30.0	40.0	50.0	60.0
Instrument response (absorbance units)	0.015	0.043	0.054	0.085	0.101	0.122

Figure 1 ***Plotted points***

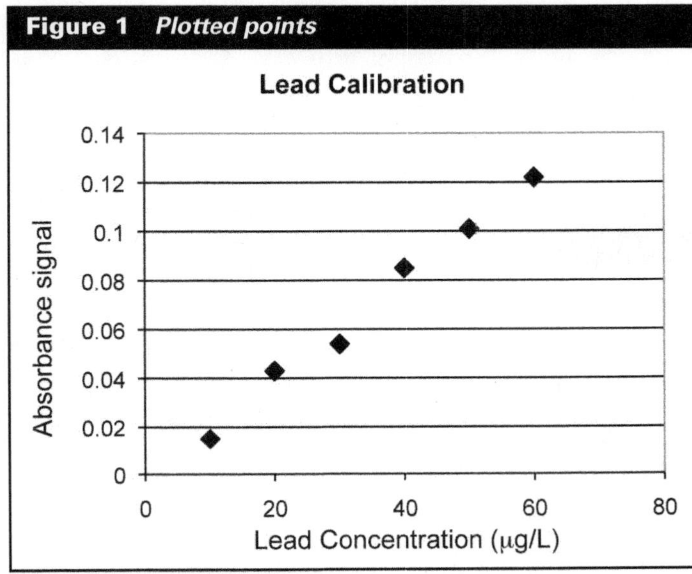

A larger version of this graph is at the end of the activity.

Key Questions

11. Theoretically, the atomic absorption signal should have a straight line relationship with the concentration of lead (Beer's Law). Does this data set seem to confirm that behavior?

12. If you were going to place a straight line on the graph, consider where it should be placed. Sketch a possible location for that line on your graph. Agree on and write down an explanation for why you decided to put it in that location.

13. Using Excel, it is possible to perform a *regression analysis,* which is a statistical procedure that determines the location of a function (e.g. in this case, a straight line) to best represent the position of the data set. Several types of output can be selected, in particular a Fitted Line Plot, a Residuals Table, and a Residuals Plot are shown below. Describe to each other what information you see here.

Figure 2 *Fitted Line* $R = 0.9951$ $R^2 = 0.9902$

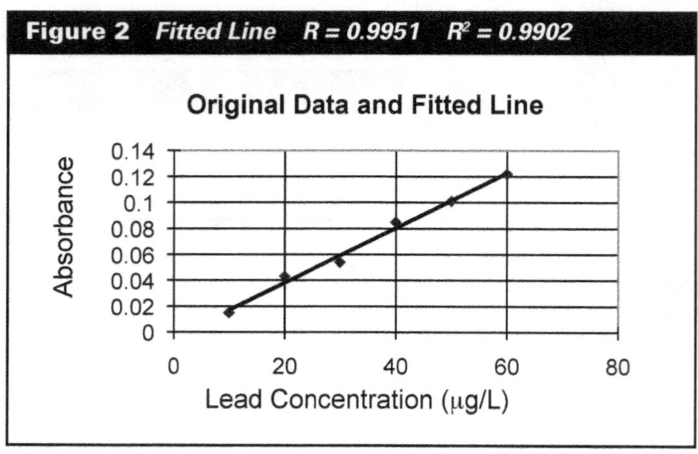

Figure 3 *Residuals Plot*

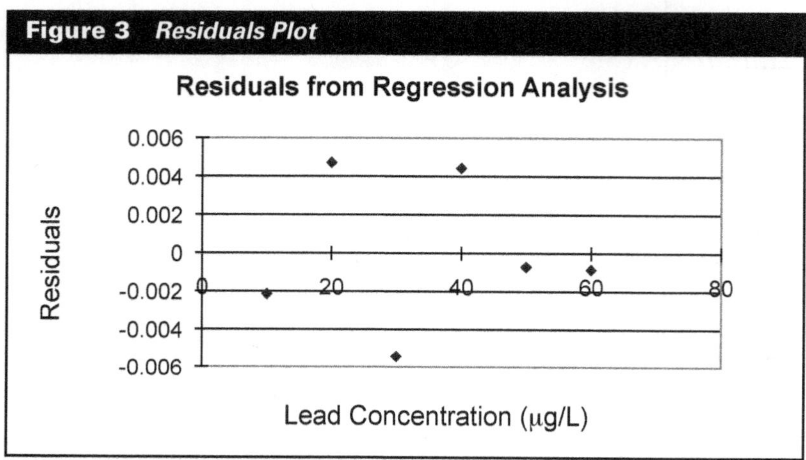

Table 3 *Residuals Table*

Observation	Predicted Y	Residuals
1	0.017143	-0.002142857
2	0.038286	0.004714286
3	0.059429	-0.005428571
4	0.080571	0.004428571
5	0.101714	-0.000714286
6	0.122857	-0.000857143

14. How well does your rough line drawn in Q12 match the location of the "fitted line" in the plot shown in Figure 2?

15. In Table 3, look for the column "Predicted Y." Verify by inspection that these data define the predicted line in the graph Figure 2.

16. Inspect Figure 3 (called a *residuals plot*). Explain what the "Residuals" axis value of zero represents. In the Residuals Table 3, compare the column of "residuals" to the "residual plot." How are these related?

17. The Excel regression analysis determines the placement of the line by the "method of least squares." Remind yourselves of the discussion you had in Q1-Q10, and look at the Residuals Plot from the Excel analysis. What do you think "least squares" means in the context of fitting a line to data?

18. Explain what steps you would have take to generate a graph similar to that in Q7-Q8 to demonstrate that the regression line is a least squares result.

19. There is an intentional relationship between the graph you constructed in Q2-Q4 and the Residuals Plot discussed in Q16. Look at them carefully and comment. Based on this comparison and the value you calculated previously for the SS residuals in Q5, guess at the value for the SS Residuals for the regression analysis.

20. Verify this planned coincidence by calculating the variance from the SS Residuals (divide SS Residuals by (n-1), the number of degrees of freedom), and compare with the variance shown in Table 1. Comment.

Consider this...

Various indicators are used to express how closely the data fit a straight line function. Two common ones are the well-named Correlation Coefficient, R, and the poorly named Coefficient of Determination, R^2. The value of R^2 has the handy property of being equal to the fraction of *explained variance*—the amount of difference in the calibration measurements accounted for by fitting the straight line. Numerically:

$$R^2 = SS_{regression} / SS_{total}$$

SS_{total} is the sum of squared differences of each value from the mean of the values.

Key Questions

21. Each person in the group should take a turn at describing what R and R^2 mean in everyday language (that their parents or friends might be able to understand). Write those down.

22. What is the range of possible values of R and R^2? What values do you think represent a "good fit" in terms of calibration? After making this decision, review the values of R and R^2 for Figure 2 to see whether your intuition is on track.

Consider this...

In designing an instrumental measurement, an important consideration is establishing the reliability of the relationship of instrumental response to analyte quantity. Regression analysis determines what this relationship is. Regression analysis can also be used to monitor the stability of the relationship. Doing so is considered an essential part of quality control.

How will you know when something is wrong? Table 4 shows three sets of calibration data obtained for the same Atomic Absorption measurement of lead on different days. Linear regression analysis is conducted for each case. Below the table is Excel output: the values of R and R^2 and the Residuals Plot.

Table 4

Lead Concentration (µg/L)	10.0	20.0	30.0	40.0	50.0	60.0
Case 1	0.011	0.041	0.059	0.077	0.092	0.100
Case 2	0.018	0.041	0.059	0.083	0.095	0.140
Case 3	0.014	0.016	0.034	0.036	0.054	0.056

Figure 4 *R = 0.9831 R^2 = 0.9664*

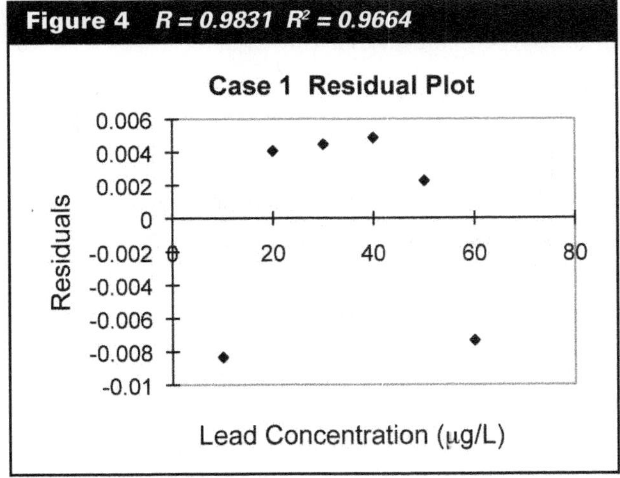

Figure 5 *R = 0.9856 R^2 = 0.9715*

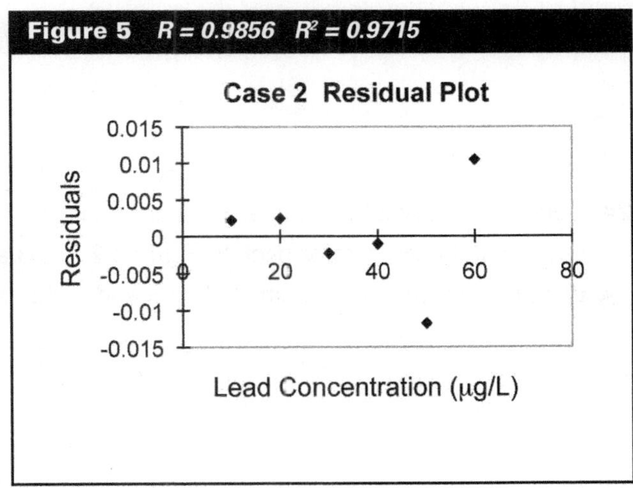

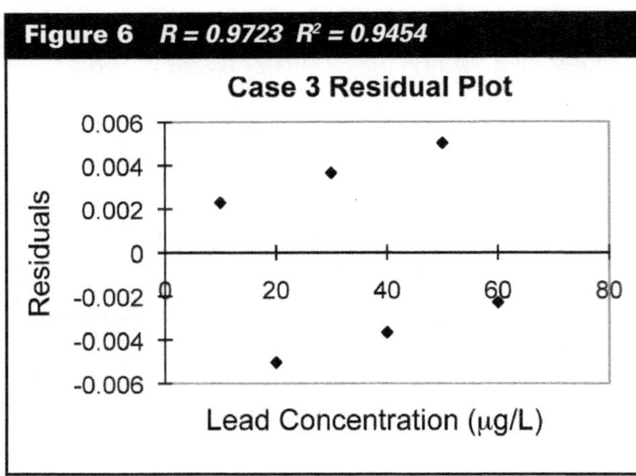

Key Questions

23. Look at the residual plots (Figures 4-6). Each group member should write a sentence describing how the points are distributed on one of the plots. Then share these descriptions with each other. Be as specific and descriptive as you can. Come to an agreement on what the "problem" seems to be in each Case. Each Case demonstrates the failure of an underlying assumption in linear regression. Write your group consensus descriptions here.

24. Comment on what the values of R or R^2 tell you (or not) about the existence of potential problems in each case. It will help to refer back to Figures 2 and 3 as a case where these "problems" do not occur. You may wish to refer to the "sensitivity" of R and R^2 to these problems.

The statistical process of regression involves some important assumptions about the underlying errors (as indicated by the residuals). In particular, if the regression accounts for all of the systematic differences among the data points, then the residuals are just showing the random noise present in the method as it is being used. If there are unaccounted-for systematic differences at work among the data, it is important to know that, because this introduces error into measurements. It is important to look for evidence regarding these possible effects. In statistics terms, this is referred to as checking for *model fit. uniform variance, normality, and independence.*

Model fit	The instrumental response should be a linear function of the analyte concentration. ("Linear" is most desired for simplicity.)
Uniform variance	The size of errors should be about the same across the measurement range.
Normality	Errors should distribute in the shape of a normal curve.
Independence	Errors should not be correlated or related to each other.

25. Decide which problem is being shown by each of the three cases. Explain why you decided that.

Case	1	2	3
Problem	____________	____________	____________

One assumption is not included in Q25. It cannot be easily tested unless one has many data points. There are statistical tests and graphs designed to check this assumption, but the small data sets that are typical in many analytical situations do not often lead to definitive conclusions. You may have completed another activity about normal (Gaussian) distributions that delves more deeply into this issue.

26. As a group, consider whether each person has had an opportunity to contribute to the discussion and whether each person feels confident about his/her understanding of linear regression.

27. Each group member should identify one specific thing he/she learned about regression analysis.

28. The group should consider what aspect of regression analysis is still not clear to them. (Perhaps the application questions will help clarify.)

Applications

29. If you were the manager of the laboratory in which the three calibrations in Table 4 were generated, suggest a course of action that might correct the observed problems for these two different situations:

(a) All three cases were from a single analyst calibrating a single instrument on three different occasions.

(b) Each case was a different analyst working on different instruments.

30. Assume that you have a severe signal drift problem (magnitude of signal gradually changes in one direction over time). What problem would this introduce if you analyze your calibration standards in sequential order of lowest to highest concentration (a very common procedural approach)? Analyzing the standards in random sequence will eliminate this source of bias. Explain why. If you do not know whether drift problems exist, as lab manager, what would your recommendation be on sequencing standards?

31. It is true that if a set of data describes a straight line with non-zero slope, then the correlation coefficient will have a value close to 1. Is the converse necessarily true? If you find a set of data where the correlation coefficient is close to 1, does that mean the data describe a straight line? Take a position and write down your response. If you decide that the correlation coefficient does not necessary tell you "the data lie on a straight line," then what else can you do to make sure you don't come to the wrong conclusion?

32. What does a linear regression do and what meaning is attached to the residuals from a regression analysis? Write down an answer as an individual. Then, discuss until you achieve a group consensus on wording, and write that down.

33. Carry out the suggestion you put forward in Q17-Q18. Split up the work among the group members and combine your results.

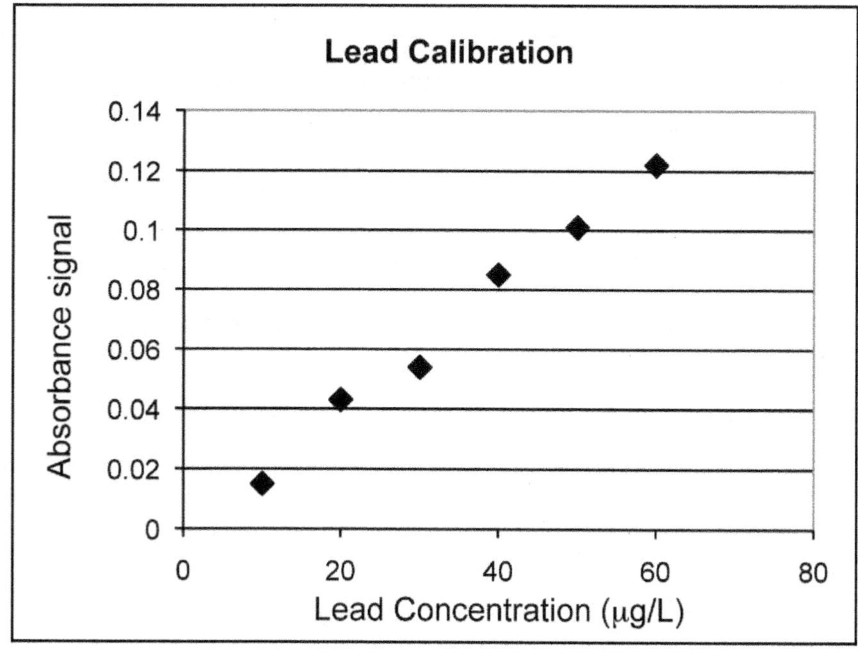

Electrochemistry: The Microscopic View of Electrochemistry

Learning Objectives

Students should be able to:

Content

- Describe microscopic-level processes occurring at electrodes and explain how these processes result in formation of the electrochemical double layer.
- Explain the origins of electrochemical potential.
- Use E° values to predict the spontaneous direction in electrochemical reactions.

Process

- Interpret pictures of electrochemical processes. (Information Processing)
- Draw and interpret microscopic pictures of electrochemical processes. (Information Processing)
- Draw and interpret graphical representations of electrochemical processes. (Critical Thinking)

Prior knowledge

- General understanding of electrostatics (e.g., charged particle behavior).
- Definitions of oxidation, reduction and standard reduction potentials.
- The use of ΔG as a predictor of spontaneous reactions.

Further Reading

- D.C. Harris, *Quantitative Chemical Analysis,* 7th Edition, 2007 W.H. Freeman: USA, Sections 14-1 through 14-3, pp. 270-9.
- D.A. Skoog, D.M. West, F.J. Holler, S.R. Crouch, Fundamentals of Analytical Chemistry, 8th Edition, 2004 Thompson Brooks/Cole: USA, Sections 18B, 18C-1 through 18C-3, pp. 496-508.
- Thomas Greenbowe's electrochemistry animation site: www.chem.iastate.edu/group/Greenbowe/sections/projectfolder/animationsindex.htm
- Özkaya, A.R., Üce, M. and Sahin, M., J. *Chem. Educ.* 2006 83(11) pp.1719-23.
- Electrochemistry Dictionary: www.electrochem.cwru.edu/ed/dict.htm
- Martins, G.F. Why the Daniell Cell Works! *J. Chem. Educ.,* 1990, 67 (6), pp. 482.

Authors

Christine Dalton and Mary Walczak

Consider this...

Figure 1(a) below contains a strip of zinc metal and a beaker containing a solution of the corresponding metal ion (e.g., $Zn(NO_3)_2(aq)$).

In Figure 1(b), the Zn metal strip is placed in the Zn^{2+} solution. The (+) and (-) signs are used to represent the electrostatic interface that forms at the metal/solution interface. These signs do not indicate the positive and negative ions in solution or free electrons in the metal electrode, but they represent the overall charge on the metal electrode and the overall charge of the solution near the metal electrode. The signs in Figure 1(b) are not meant to be absolute values, but rather relative to one another.

Figure 1(c) is the microscopic view of the interface between the surface of the electrode and the metal ion solution within which the metal electrode is placed, not showing NO_3- and H_2O for simplicity. Electrons within the metal surface are delocalized.

Figure 1 ***Microscopic Processes at Metal Surfaces: Zinc metal and $Zn(NO_3)_2$ solution.***

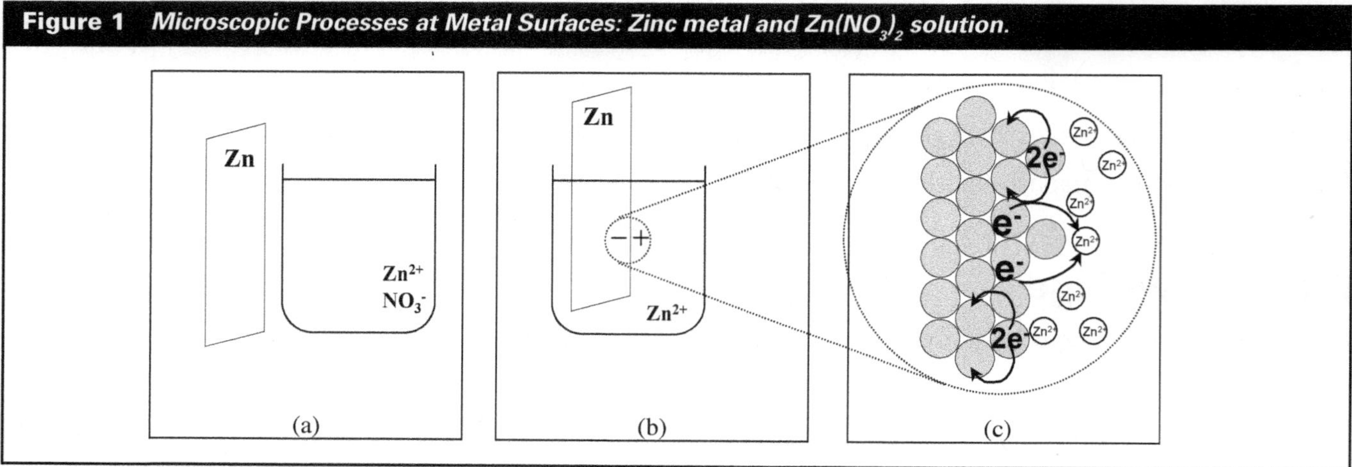

Key Questions

1. Examine the three panels in Figure 1. Classify each panel as representative of the situation *before* or *after* immersion.

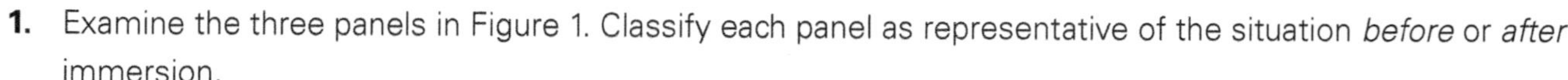

(a) (b) (c)

What is the relationship between panels (b) and (c)?

2. A Zn^{2+} ion collides with the electrode and gains 2 e-. What is the reaction for this event? Is this an oxidation or a reduction?

3. A Zn atom on the electrode loses 2 e- which join the collection of delocalized electrons on the metal. What is the reaction for this event? Is this an oxidation or a reduction?

4. Examine figure 1(c) closely. Three microscopic processes are shown. One is characterized by movement of delocalized electrons to a zinc ion. Two others show the loss of electrons by a Zn atom. Label the oxidation and reduction processes in figure 1c.

Key Questions

A strip of zinc metal is placed in a solution of $Zn(NO_3)_2$ and equilibrium is established. Note: Nitrate ions and water molecules are omitted for clarity.

Figure 2 ***Predominant Microscopic Processes at Metal Surfaces: Zinc metal in $Zn(NO_3)_2$ solution (a) when initially immersed and (b) once equilibrium is reached.***

(a) Before equilibrium

(b) At equilibrium

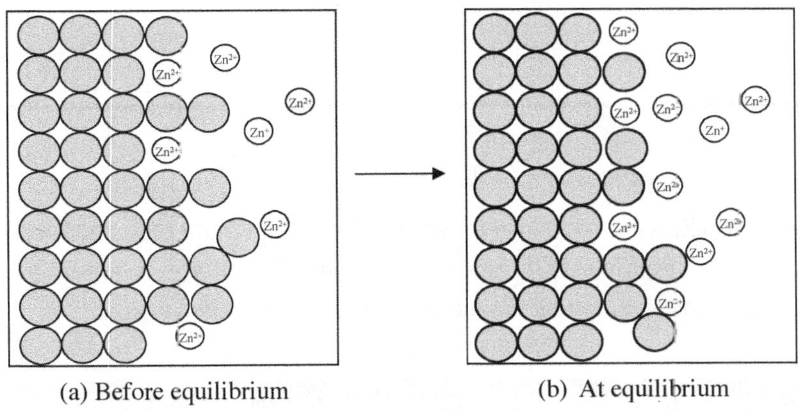

5. Draw arrows to show where the electrons are going as equilibrium is reached. Confirm that the drawing shows that *oxidation* is the favored reaction.

6. Describe the microscopic processes that occur when a strip of zinc metal is placed in a solution of $Zn(NO_3)_2$.

7. Which of the following reactions occurs to a greater extent? Explain your reasoning.

$Zn^{2+}(aq) + 2e^- \rightarrow Zn(s)$ $\qquad$ $Zn(s) \rightarrow Zn^{2+}(aq) + 2e^-$

8. Divide the following two questions among group members. Circle your answer to the following questions. Come to consensus on your answers.

   Based on your answer to Q5-Q7, is the charge at the surface of the Zn electrode:

   Positive Negative Neutral?

   Explain your reasoning.

   Based on your answer to Q5-Q7, is the charge in the solution in the vicinity near the electrode:

   Positive Negative Neutral?

   Explain your reasoning.

9. Verify your answers to Q8 by looking at Figure 1b.

Consider this...

The following figure shows the Zn electrode immersed in a $Zn(NO_3)_2$ solution at equilibrium. Note: Solvent molecules have been removed for simplicity. Suppose you have a tiny charge measuring device that you can position at various locations in the solution to measure the charge at a single point.

Figure 3 ***Schematic diagram of the surface of Zinc metal in $Zn(NO_3)_2$ solution at equilibrium.***

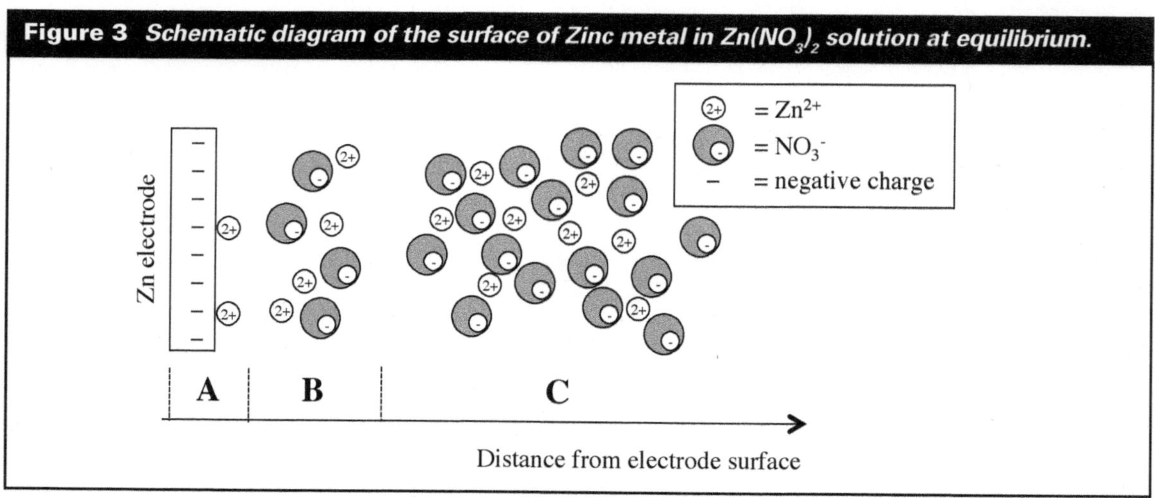

10. What is the overall charge in each of the following regions of the figure?

Region A	Positive	Neutral	Negative
Region B	Positive	Neutral	Negative
Region C	Positive	Neutral	Negative
Regions A and B	Positive	Neutral	Negative
Regions B and C	Positive	Neutral	Negative

11. How does the charge measured by your device change if the probe is moved from point A to point B in the figure?

12. What happens to the local charge when you move the probe from point B to point C?

13. On the following graph of charge versus distance from the electrode, label the location of points A, B, and C from Figure 3. As a group, use the graph to explain what happens to the charge as the distance from the electrode increases.

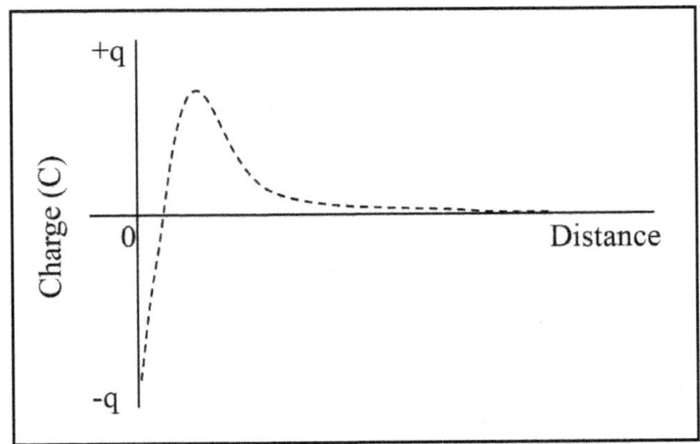

The electrical double layer (EDL) occurs at the interface between an electrode and the solution. It is comprised of adjacent positive and negative regions, as shown in Figure 3. The inner layer (A) has charged particles directly adsorbed onto the surface of the electrode, while the outer layer (B) has ions electrostatically attracted to the adsorbed charged particles. This second layer is called the diffuse layer. The net electric charge in the diffuse layer is equal in magnitude to the net electric charge on the surface but with the opposite charge, leading to a net electrically neutral EDL.

14. Indicate the location of the EDL on Figure 3 and on the graph in Q13.

Consider this...

Figure 4(a) below contains a strip of lead metal and a beaker containing a solution of the corresponding metal ion (e.g., $Pb(NO_3)_2$*(aq)*).

In Figure 4(b), the Pb strip is placed in the Pb^{2+} solution.

Figure 4c is the microscopic view of the interface between the surface of the electrode and the metal ion solution within which the metal electrode is placed, not showing NO_{3-} and H_2O for simplicity.

Figure 4 ***Microscopic Processes at Metal Surfaces: Lead metal and $Pb(NO_3)_2$ solution.***

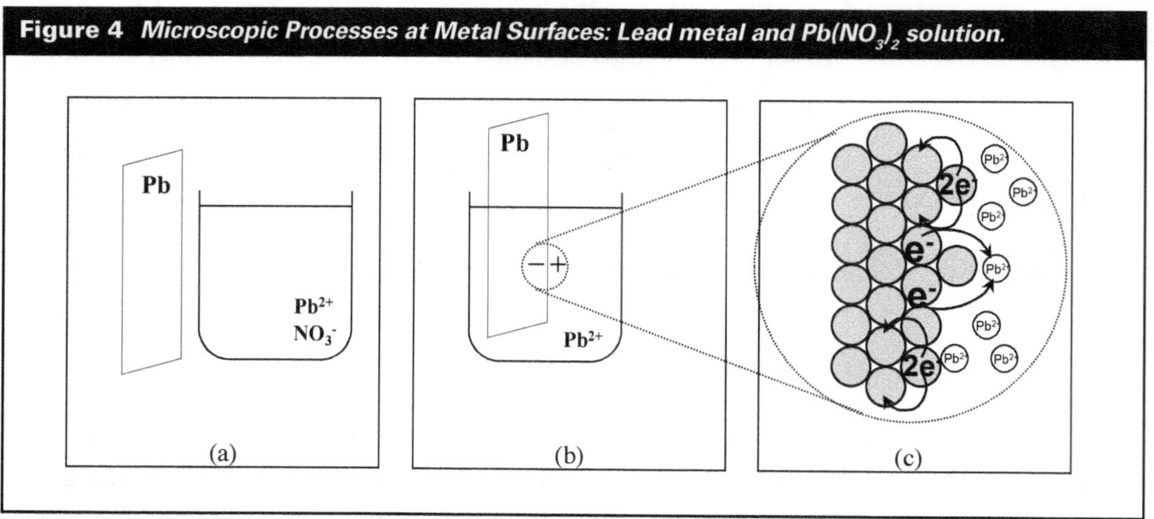

Key Questions

15. Divide into pairs to answer the following question: According to figure 4(c) which of the following reactions occur immediately after a strip of lead metal is placed into a solution of $Pb(NO_3)_2$*(aq)*?

I. $Pb^{2+}(aq) + 2e^- \rightarrow Pb(s)$

II. $Pb(s) \rightarrow Pb^{2+}(aq) + 2e^-$

A) I only B) II only C) I and II D) none

Share your answer with the other pair. Come to consensus and explain the rationale behind your choice.

16. Using the figures below as a starting point, draw the microscopic processes that occur when a strip of lead metal is placed in a solution of $Pb(NO_3)_2$. Draw in the arrows to show where the electrons are going. Note: Nitrate is omitted for clarity.

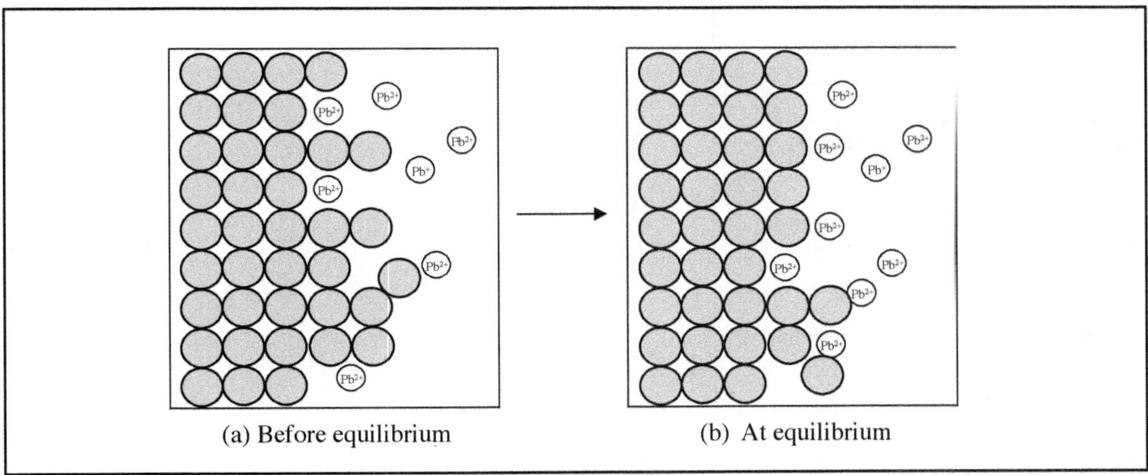

17. Now that you have explored a Zn/Zn^{2+} and Pb/Pb^{2+} metal-solution interface, list the differences you have discovered between the two interfaces.

18. Using Figure 3 as a model, draw the charge distribution (electrical double layer) AT the surface of the electrode (point A), farther away from the electrode (point B), and far away from the electrode (point C) for the Pb/Pb^{2+} system. Label the inner layer, the diffuse (outer) layer and the bulk solution. Make sure your drawing shows the differences you identified in Q17.

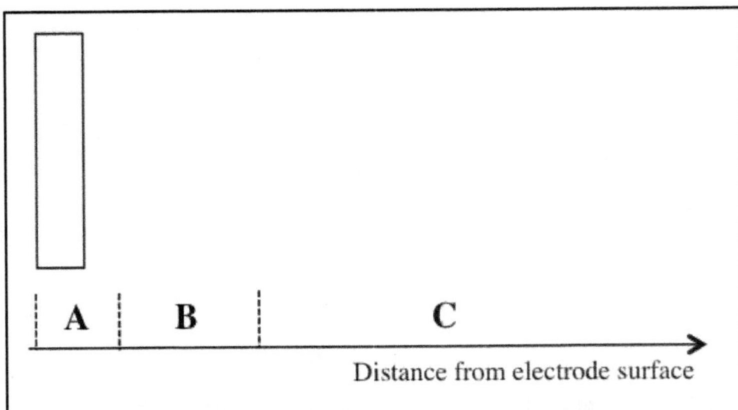

19. Compare Zn and Pb in the following graph of charge versus distance from the electrode.

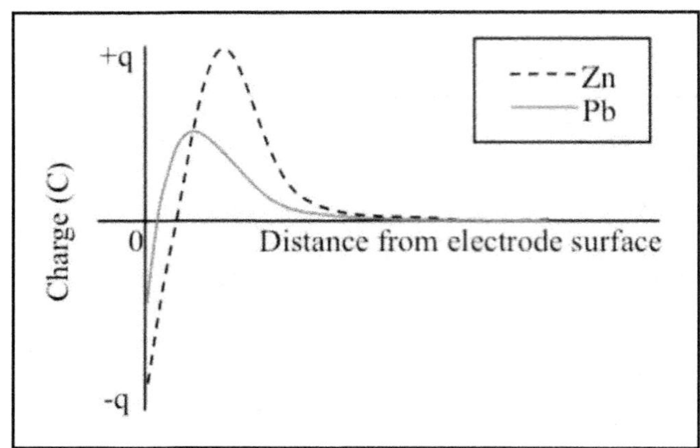

(a) Which electrode has a double layer that extends farther into solution from the metal surface?

(b) Which electrode has the larger residual negative charge at the metal surface?

(c) What evidence from the graph led to your answers in (a) and (b)?

20. The following drawings of double layer region represent the charge distribution for a Zn and for a Pb electrode immersed in a solution of its nitrate salt. Identify which picture represents Zn and which picture illustrates Pb. List two criteria you used in arriving at your answer.

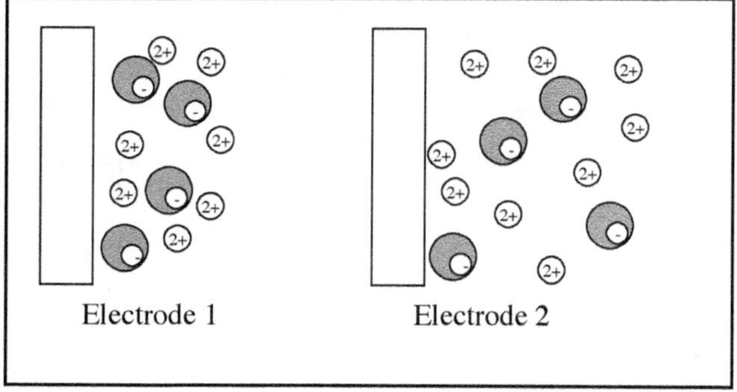

21. Using the information in the previous two questions, list the similarities and differences in the charge distribution for Zn and Pb electrodes in solution. Compare your answers with another group.

22. In your own words, define the electrical double layer for a metal placed into a solution containing the metal cation (i.e., Zn/Zn^{2+} or Pb/Pb^{2+}).

Consider this...

The following energy level diagrams show a three-step process for converting a metal atom into a hydrated divalent cation.

Figure 5 *Enthalpy changes for converting a metal atom into a hydrated divalent cation.*

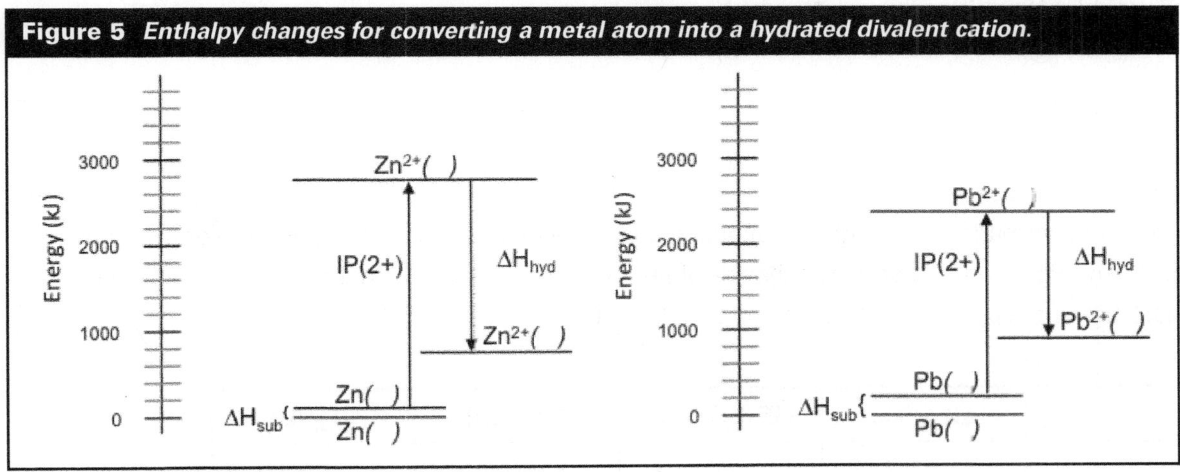

Enthalpy is a state function. The change in enthalpy between the initial and final species is the same, regardless of path. As such, the path from one chemical species to another does not matter.

23. The first step of the process, annotated as ΔH_{sub} in the figure, is the sublimation of zinc atoms. The value of ΔH_{sub} is 130 kJ for Zn and 197 kJ for Pb. On the figure, write in the states of matter for the Zn in the sublimation of zinc. Write the chemical reaction that occurs for the sublimation step.

24. The second step of the process, annotated as IP(2+) in the figure, is the ionization of the gas phase atom to the divalent cation. The value of IP(2+) is 2641 kJ for Zn and 2168 kJ for Pb. On the figure, write in the state of matter for the zinc divalent cation after gas phase ionization. Write the chemical reaction that occurs during the gas phase ionization of zinc atoms.

25. The third step of the process, annotated as ΔH_{hyd} in the figure, is hydration of the divalent cation. The value of ΔH_{hyd} is -2009 kJ for Zn and -1486 kJ for Pb. On the figure, write in the state of matter for the divalent cation after hydration. Write the chemical reaction that occurs during the hydration of divalent zinc cations.

26. Write the overall chemical reaction for Zn that is represented in the above figure. Verify that you have the correct reaction by comparing your answer to the figure legend.

27. Repeat Q23-Q26 for Pb.

28. The following table summarizes the enthalpy of sublimation, the ionization potential, and the enthalpy of hydration for Zn and Pb. Calculate enthalpy change of reaction for Zn and Pb and add your answers to the table. Compare the enthalpy of reaction that you calculated for Zn and Pb.

Metal	ΔH_{sub} (kJ)	IE (2+) (kJ)	ΔH_{hyd} (kJ)	ΔH_{rxn} (kJ)
Zn	130	2641	−2009	
Pb	197	2168	−1486	

29. Based on your answer to Q19 about the electrical double layer, which metal is easier to oxidize, Zn or Pb?

30. Do the relative magnitudes of the ΔH_{rxn} values calculated in Q28 support or refute the conclusions from the trend from Q19? Explain your reasoning.

31. A piece of Zn metal is placed into a solution of $Pb(NO_3)_2$. The predicted reactions are:

$Zn(s) \rightarrow Zn^{2+}(aq) + 2e^-$

$Pb^{2+}(aq) + 2e^- \rightarrow Pb(s)$

Write the overall oxidation-reduction reaction. Which metal is oxidized and which is reduced?

32. Calculate the enthalpy change (ΔH) for the *overall* oxidation-reduction reaction written above. What is ΔG for this reaction? Is this reaction spontaneous?

We know from earlier study of thermodynamics that a reaction is spontaneous if the change in free energy for the process is negative (i.e., $\Delta G = \Delta H - T\Delta S < 0$). For the redox reactions under consideration here (i.e., $A(s) + B^{2+}(aq) \rightleftharpoons A^{2+}(aq) + B(s)$), the entropy change is small, so the reaction is under control of the enthalpy change ($\Delta G \cong \Delta H$).

Consider this...

By convention, the driving force for a particular electrochemical reaction is quantified by the Standard Reduction Potential (E°). Each redox pair (e.g., Zn(s) and $Zn^{2+}(aq)$) has its own intrinsic standard reduction potential; the more positive the potential, the greater the species' affinity for electrons and tendency to be reduced. A more positive E° means there is a greater tendency for reduction to occur. Standard reduction potentials – as shown below in Table 1 – are the potentials for the specific situation in which all species are present *in the standard state,* that is at 1 M concentration for solution species and 1 bar pressure for gases. The small ° symbol indicates the value is a standard state value. Tables listing E° values at 25°C can be found in many textbooks and reference books.

Table 1 ***Standard Reduction Potentials***

Reaction	E°(Volts)
$Ag^+ + e^- \rightleftharpoons Ag(s)$	0.799
$Cu^{2+} + 2e^- \rightleftharpoons Cu(s)$	0.337
$2H^+ + 2e^- \rightleftharpoons H_2(g)$	0.000
$Pb^{2+} + 2e^- \rightleftharpoons Pb(s)$	−0.126
$Cd^{2+} + 2e^- \rightleftharpoons Cd(s)$	−0.403
$Zn^{2+} + 2e^- \rightleftharpoons Zn(s)$	−0.763

Key Questions

33. Which cation, Zn^{2+} or Pb^{2+}, is easier to reduce? Explain how you decided.

34. Based on your answer to Q29 where the energetics of the oxidation process were explored, which metal, Zn or Pb is easier to oxidize.

35. Are your answers to Q33 and Q34 consistent with one another?

36. Does the microscopic view of the interface (Q19) support or refute the trend in the standard reduction potentials (Q31)?

37. Is Cd^{2+} easier to reduce than Zn^{2+}? Than Pb^{2+}?

Is Cd metal easier to oxidize than Zn? Than Pb?

Consider this...

Reduction reactions, such as those listed in Table 1, do not occur independently. The electrons consumed in the reduction reaction must be generated in a coupled oxidation reaction. Oxidation-reduction reactions are based on the combination of two half-reactions (an oxidation and a reduction). Since potentials can only be measured as differences, a zero point is arbitrarily assigned to the reaction $2H^+ + 2e^- \rightleftharpoons H_2(g)$, which occurs at the standard hydrogen electrode (SHE). Therefore, E° values in Table 1 are relative to the standard hydrogen electrode, which is defined as having E°equal to 0 Volts.

Combining two half-reactions can result in construction of an electrochemical cell. The difference in free energy ΔG° between the products and reactants $G^\circ_{prod} - G^\circ_{react}$ determines whether the reaction is favored to proceed as written.

$Zn(s) \rightleftharpoons Zn^{2+}(aq) + 2e^-$
$Pb^{2+}(aq) + 2e^- \rightleftharpoons Pb(s)$
$Pb^{2+}(aq) + Zn(s) \rightleftharpoons Zn^{2+}(aq) + Pb(s)$

The potential difference measured between the two electrodes, E°_{cell}, is also a measure of the extent to which the reaction is favored to proceed as written. The value of E°_{cell} is determined using the standard reduction potentials: $E^\circ_{cell} = E^\circ_{red} - E^\circ_{ox}$, where E°_{red} is the standard reduction potential value (E°) for the reaction undergoing reduction and E°_{ox} is the E° value for the reaction undergoing *oxidation*. Note that the calculation of E°_{cell} is a *difference*, that is a subtraction.

There are other ways of calculating E°_{cell} that you may have learned in earlier chemistry courses. We chose to utilize this method here as it strictly emphasizes the potential *difference* between two electrodes.

Key Questions

38. Using the values in Table 1, calculate E°_{cell} for the oxidation-reduction reaction:

$Pb^{2+}(aq) + Zn(s) \rightleftharpoons Zn^{2+}(aq) + Pb(s)$

39. Based on your answer to Q32 above, is this reaction spontaneous as written?

Recall that free energy (ΔG°) and E°_{cell} are directly related through the equation $\Delta G^\circ = -nFE^\circ_{cell}$, where n is the number of electrons exchanged in the redox reaction and F is Faraday's constant (96485 C/mole e-). Thus, the cell potential E°_{cell} for an oxidation-reduction reaction can be calculated and used to assess the spontaneity of the reaction.

40. Using E°_{cell}, calculate ΔG° for the reaction in Q38.

41. Compare your answers in Q32 and Q40.

42. Write the reverse reaction (i.e., reacting Pb(s) with Zn^{2+}) to the overall oxidation-reduction reaction in Q38. Calculate the E°_{cell} for the reverse reaction. Is the reverse reaction spontaneous?

43. Imagine two situations. In one beaker, a piece of Cu metal is placed into an aqueous solution of $AgNO_3$, while in the second beaker, a piece of Ag metal is placed into an aqueous solution of $Cu(NO_3)_2$.

a) Divide the two beakers among group members. Each subgroup should write the *overall* oxidation-reduction reaction for the reaction occurring in their assigned beaker. Come to consensus on the two reactions occurring in the two beakers.

b) Which reaction in part a is spontaneous as written?

c) Which of the two metal ions, Cu^{2+} or Ag^+, is more likely to be reduced? Explain your reasoning.

44. How has this activity extended your conceptual picture of electrochemical processes/reactions over what you attained in general chemistry?

45. What skill(s) did you gain in interpreting microscopic representations of electrochemical cells during this activity?

Applications

46. A drop of Hg metal is placed in a 1M $Hg_2(NO_3)_2$ solution. Describe what happens on a microscopic scale to the Hg metal and the Hg_2^{2+} ions. Draw a microscopic picture showing what happens and sketch a graph of charge vs. distance for this metal-solution interface. *Note: at this interface, reduction is the favored process.*

47. Considering Figures 2 and 3, in your own words describe the microscopic-level processes in terms of equilibrium that occurs in a beaker containing a metal electrode and its corresponding metal ion in solution.

48. Write the electrochemical equilibrium reaction when a strip of Cd metal is placed into a 1M $Cd(NO_3)_2$ solution. Is oxidation or reduction the favored process? What is the charge at the surface of the Cd electrode? What is the charge in the solution in the vicinity of the Cd electrode? Sketch a graph of charge vs. distance for this metal-solution interface. *Hint: Where does the Cd half-reaction fall in Table 1?*

49. If the standard reduction potential (E°) is +1.507V for the following reaction, is reduction or oxidation favored?

$MnO_4^- + 8H^+ + 5e^- \rightleftharpoons Mn^{2+} + 4H_2O$

50. The charge at the surface of a metal electrode immersed in a salt solution of the corresponding metal nitrate as a function of distance from the electrode surface is shown below. Draw a microscopic-level picture of the electrode surface and the solution in the vicinity near the electrode. Is the metal or metal ion side of the equilibrium favored?

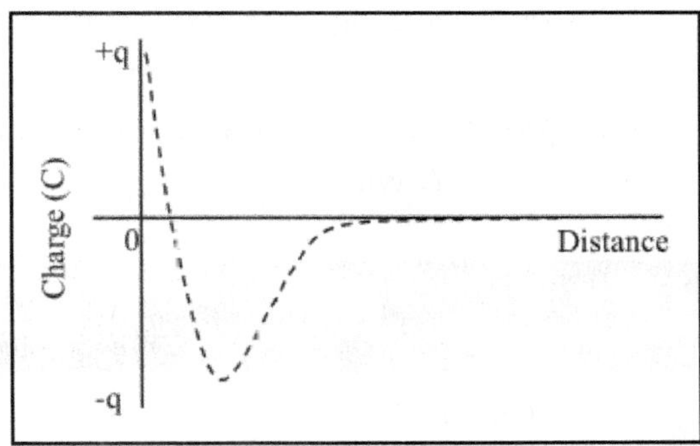

51. Four metals, A, B, C, and D, exhibit the following properties:

a) Only A and C react with 1.0 M HCl to give $H_2(g)$.

b) When C is added to solutions of the ions of the other metals, metallic B, D, and A are formed.

c) Metal D reduces B^{n+} to give metallic B and D^{n+}.

Based on the information above, arrange the four metals in order of **increasing** tendency toward oxidation.

52. The objective of this problem is to identify an unknown metal. The metal may be Fe, Mg, Ni, Pb, Sn or Zn. A microscopic-level picture of the surface of this metal when placed in a solution containing the metal nitrate is shown at the right. Water molecules are omitted for clarity. This metal is also placed into solutions of all the metal nitrates corresponding to the possible metals. Reaction occurs when the metal is placed in $Pb(NO_3)_2$ and $Sn(NO_3)_2$ solution only. Identify the mystery metal.

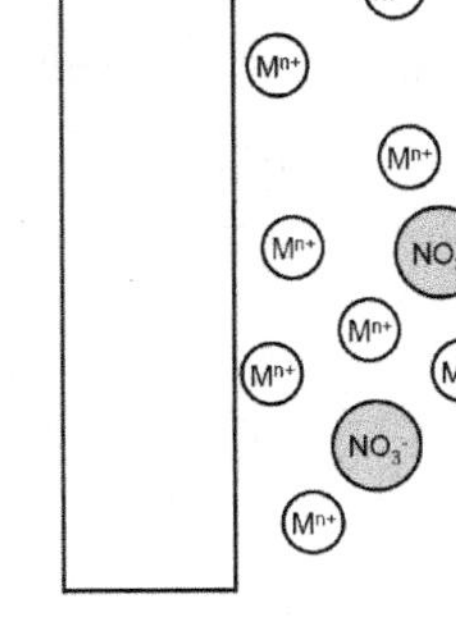

53. Thermodynamic data for the formation of $Cu^{2+}(aq)$ from Cu(s) appears in the table below. From this information, draw an energy diagram showing the three steps associated with the values in the table. What is the enthalpy of reaction for this overall reaction? Comparing the results for Cu with those for Zn and Pb (Q28), is oxidation favored in Cu to a greater or lesser extent than in Zn and Pb?

Reaction	ΔH_{sub}(kJ)	IE (1+ or 2+) (kJ)	ΔH_{hyd}(kJ)	ΔH_{rxn}(kJ)
$Cu^{2}+ + 2e- \rightleftharpoons Cu(s)$	338	2703	-2061	

54. The electrochemical cell shown here has a zinc electrode and a copper electrode. The aqueous solution into which both electrodes are placed originally contains 1 M $Zn(NO_3)_2$ and 1 M $Cu(NO_3)_2$. Sketch the interface between each electrode and the solution similar to Figure 2. Include both a "before" and "after" picture for the Zn and the Cu electrode. What reactions will occur at each metal surface?

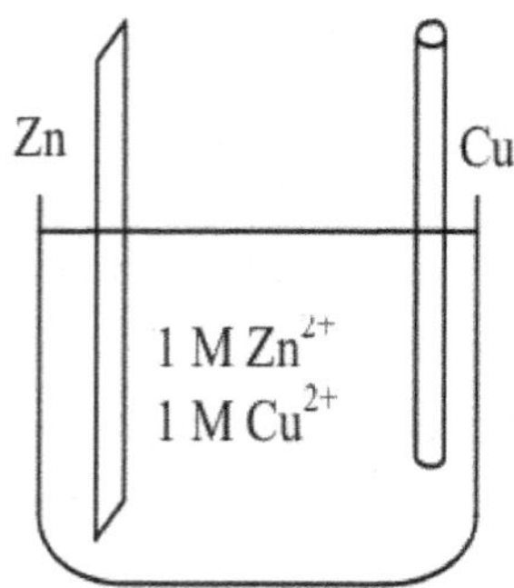

Electrochemical cell with a zinc electrode and a copper electrode placed into a solution of $Zn(NO_3)_2$ and $Cu(NO_3)_2$.

55. A piece of silver metal is placed into a solution containing $Fe(NO_3)_3$. Does a spontaneous reaction occur? Sketch the interface between the silver metal surface and the solution.

56. Electromotive force (emf) is the work done by a source on an electrical charge. Charges are moved from lower electrical potential to higher electrical potential. A capacitor can be used to demonstrate emf because a capacitor is designed to store opposite charges on two parallel plates. After a battery is used to build up charge on the two plates (charge the capacitor), electrons can be made to flow from the negatively charged plate to the positively charged plate by connecting the two plates with a wire.

Figure 6 ***A charged capacitor at different levels of charging with a positive test charge between the plates to assess the effect of emf.***

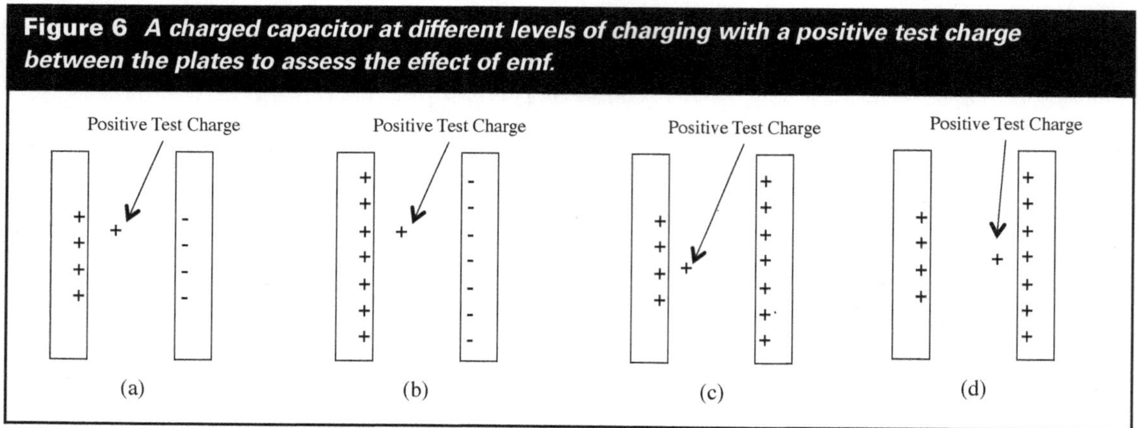

Figure 6(a) and 6(b) demonstrate the emf on a positive test charge residing between the two plates of a capacitor. In figure 6(a), there is less charge build-up on the plates than in Figure 6(b), and in Figure 6(c), the charge is the same on the two plates with different magnitudes of charge on opposite plates.

a. In Figure 6(a), which direction does the positive test charge want to travel?

To the positive plate or to the negative plate

b. In Figure 6(b), which direction does the positive test charge want to travel?

To the positive plate or to the negative plate

c. Compare figures 6(a) and 6(b).

i. In which figure (a) or (b) is the electrostatic attraction (emf) on the positive test charge greater? Why?

ii. In which figure (a) or (b) does the positive test charge require more energy to move toward the positive electrode? How does this compare to your answer in part a?

iii. In which figure (a) or (b) would more energy be released if the positive test charge were allowed to move in the direction it moves spontaneously? Why?

d. List the similarities between figures 6(a) and 6(b) with figure 6(c) and 6(d). List the differences between the same pairs of figures.

e. In Figures 6(c) and 6(d), which direction does the positive test charge want to move? Explain your reasoning. To the left or to the right

f. In which figure (c) or (d) would more energy be released if the positive test charge were allowed to move in the direction it moves spontaneously? Why?

g. Does the positive test charge experience a higher potential in figure 6(c) or 6(d)? Explain your reasoning.

h. Is the emf on the positive test charge in figure 6(c) larger or smaller than in figure 6(b)?

57. Suppose the charge vs. distance graph for a particular metal immersed in an electrolyte solution (ionic strength=0.050 M) has the form shown in the figure. How will the curve change if the ionic strength of the solution is dropped to 0.010 M? What if it is increased to 0.10 M? Explain the rationale behind your conclusions.

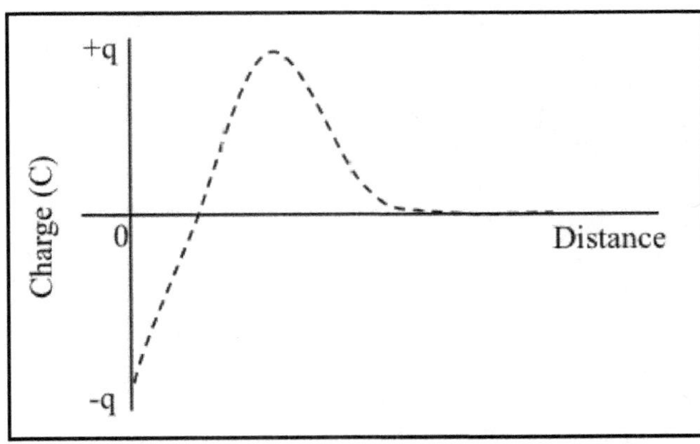

Electrochemistry: Calculating Cell Potentials

Learning Objectives

Students should be able to:

Content

- Apply the criteria for an electrochemical cell in the standard state to determine if a cell is initially in the standard state.
- Predict how the ion concentrations in an electrochemical cell will change as an electrochemical cell circuit is completed and allowed to run.
- Determine E_{cell} for a given electrochemical cell and decide whether the reaction is spontaneous in the forward or reverse direction.

Process

- Interpret drawings of electrochemical cells. (Information Processing)
- Compare and contrast different electrochemical cells. (Critical Thinking)

Prior knowledge

- Definitions of oxidation, reduction, anode, cathode, voltmeter, etc.
- Identifying oxidation and reduction processes.
- Identifying chemical reactions occurring at each electrode in an electrochemical cell.
- Calculating the cell potential for a simple electrochemical cell under standard state conditions.
- Determining if an oxidation-reduction reaction is spontaneous.

Further Reading

- Harris, D.C. 2010. *Quantitative Chemical Analysis,* 8th Edition, W.H. Freeman: USA, Sections 14-1 through 14-3, pp.309-14.
- Skoog, D.A., D.M. West, F.J. Holler, S.R. Crouch, 2004. Fundamentals of Analytical Chemistry, 8th Edition, Thompson Brooks/Cole: USA, Sections 18B, 18C-1 through 18C-3, p.496-508.

Notes

- Table 1 provided at the end of the activity is to be used for Questions 17, 18 and 25. It is helpful to keep the table separate from the activity to add information to it.
- Table 2 provided at the end of the activity is to be used for Questions 32, 33, and 37.
- A table of standard reduction potentials (E°) is provided at the end of the activity in Q50 for when E° is needed.

Author

Christine Dalton and Mary Walczak

Section 1: Determining if an electrochemical cell is in the standard state

Consider this...

In thermodynamics, the *Standard State* of a material is a reference point used to calculate the material's properties. A gas in the standard state has a pressure of 1 bar and is at 25°C, and the standard state of solids and liquids is the pure form at 1 bar and 25°C. Solutions are in the standard state when they have an activity of 1 M at 25°C. Remember A = [M]γ, where A is the activity and γ is the activity coefficient. In analytical electrochemistry, concentrations of analyte are often in the millimolar range, but supporting electrolyte concentrations may be high enough for activity coefficients to deviate from 1. For purposes of this activity, we will make the approximation that activities are equal to concentrations.

Figure 1 ***Electrochemical apparatuses with varying concentrations and connections. All of these apparatuses contain a zinc electrode and a silver electrode.***

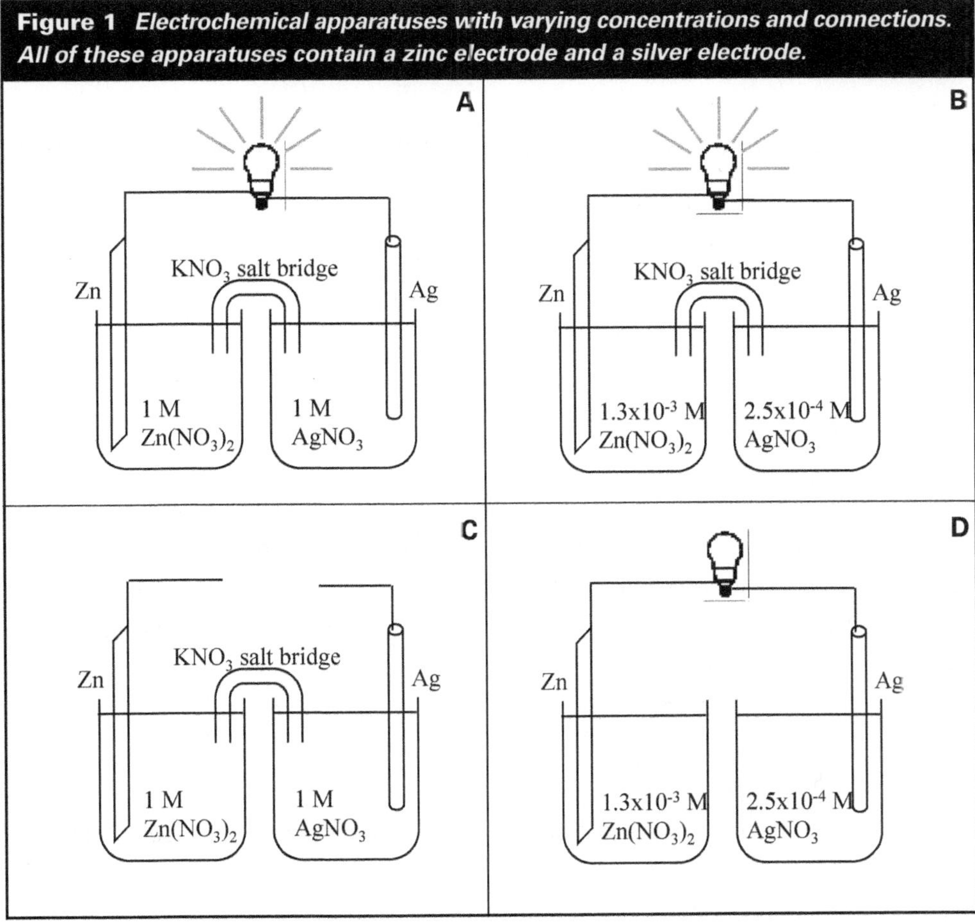

Key Questions

1. What conditions must exist for an electrochemical apparatus to be considered in the "standard state?"

2. Circle the electrochemical apparatuses in Figure 1 that meet the conditions of the standard state.

3. Figure 1 illustrates four electrochemical apparatuses. List the components that appear in these electrochemical apparatuses.

4. An electrochemical cell is distinguished from an "apparatus" in that a cell requires a complete electrical circuit. Place a star next to the apparatuses in Figure 1 that have complete electrical circuits and are therefore electrochemical cells. Compare your annotated Figure 1 with another group and resolve any differences.

5. The electrochemical "apparatuses" NOT starred in Q4 are not electrochemical cells. Explain for each why they are not cells.

6. Consider electrochemical cell A in Figure 1 (reproduced here). Given that the forward reaction in Figure 1A is spontaneous for this electrochemical cell:

Figure 1A ***Electrochemical cell 1A with the spontaneous reaction:*** $Zn(s) + 2Ag^{+}(aq) \rightleftharpoons 2Ag(s) + Zn^{2+}(aq)$

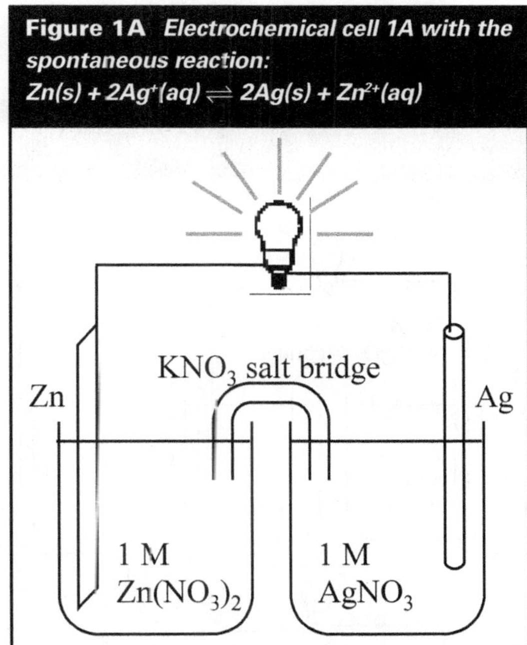

- Write the two half-cells occurring in this cell.

- Label the anode (where oxidation takes place) and cathode (where reduction takes place).

- Show the direction of spontaneous current flow in the cell.

Section 2: Ion concentrations in electrochemical cells

Consider this...

Figure 2 *Electrochemical apparatuses with varying concentrations and connections. Apparatuses A-D are the same as Figure 1.*

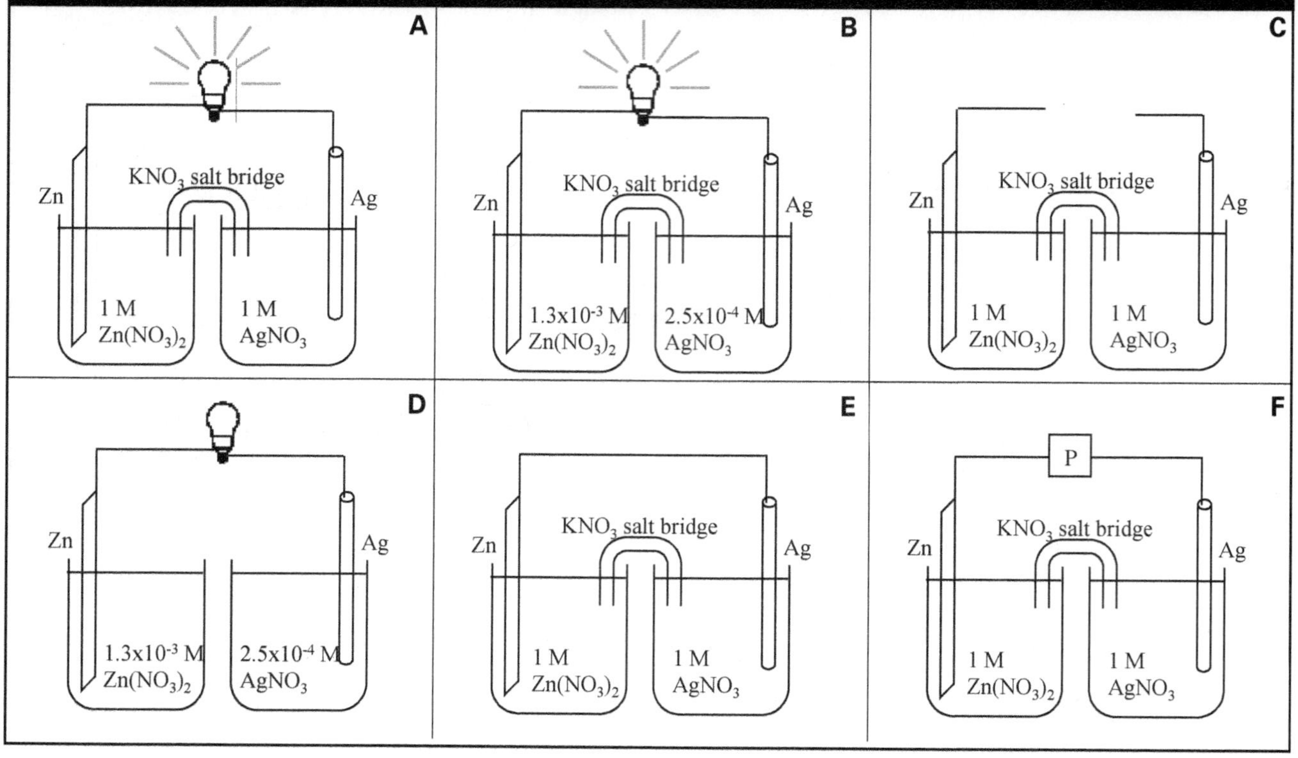

A
Zn
KNO_3 salt bridge
Ag
1 M $Zn(NO_3)_2$
1 M $AgNO_3$

B
Zn
KNO_3 salt bridge
Ag
$1.3x10^{-3}$ M $Zn(NO_3)_2$
$2.5x10^{-4}$ M $AgNO_3$

C
Zn
KNO_3 salt bridge
Ag
1 M $Zn(NO_3)_2$
1 M $AgNO_3$

D
Zn
Ag
$1.3x10^{-3}$ M $Zn(NO_3)_2$
$2.5x10^{-4}$ M $AgNO_3$

E
Zn
KNO_3 salt bridge
Ag
1 M $Zn(NO_3)_2$
1 M $AgNO_3$

F
P
Zn
KNO_3 salt bridge
Ag
1 M $Zn(NO_3)_2$
1 M $AgNO_3$

Key Questions

7. Consider electrochemical cell A in Figure 2. Note that the light bulb in the external circuit is illuminated. List any observations that would lead you to conclude that an electrochemical reaction occurs in this cell.

8. Consider electrochemical cell E in Figure 2. What observations lead you to conclude whether or not an electrochemical reaction occurs in this cell? Check your answer with another group.

9. Consider cells A and E only. Both of these cells are shown under conditions where the forward reaction (written in Q6) is spontaneous. At what point does the spontaneous electrochemical reaction occurring in cells A and E stop?

Apparatus F in Figure 2 contains a potentiometer, labeled "P." Today, multimeters, an example of which is shown at the right, have largely replaced potentiometers in chemistry laboratories. While multimeters can operate in several modes (e.g., measuring current, resistance, AC voltage or DC voltage), when DC voltage is selected it functions as a potentiometer—a high resistance voltmeter—and measures the potential (voltage) difference between two points. In the case of these cells it is the potential difference between the two electrodes.

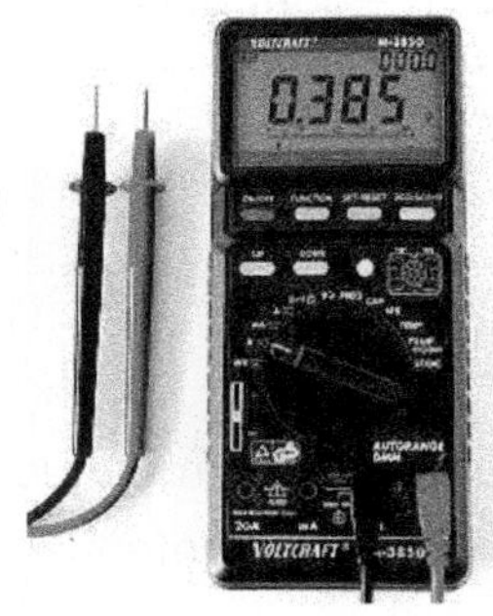

http://upload.wikimedia.org/wikipedia/commons/thumb/a/a6/Digital_Multimeter_Aka.jpg/220px-Digital_Multimeter_Aka.jpg

10. What does "high resistance" mean? Resistance to what?

11. Recall the requirements for an apparatus to be an electrochemical cell. Place a star next to the apparatuses in Figure 2 that meet the requirements and are therefore electrochemical cells. Compare your annotated Figure 2 with another group and resolve any differences.

12. Consider electrochemical apparatus F in Figure 2. Does current flow in this electrochemical cell?

13. In electrochemical apparatus F, the electrochemical reaction does not proceed. Which of the following explains why the reaction does not proceed? Justify your answer.

(a) The electrochemical circuit is not complete.

(b) The potentiometer does not allow current to flow.

(c) The forward reaction is not spontaneous.

14. Consider the electrochemical cells at the right with a potentiometer and a light bulb in series. The light bulb does not illuminate in the cell on the left. What is the most likely explanation for this behavior?

(a) The circuit is incomplete.

(b) The resistance is too high.

(c) The bulb is burned out.

Figure 3 ***Two electrochemical cells with potentiometer and light bulb in series.***

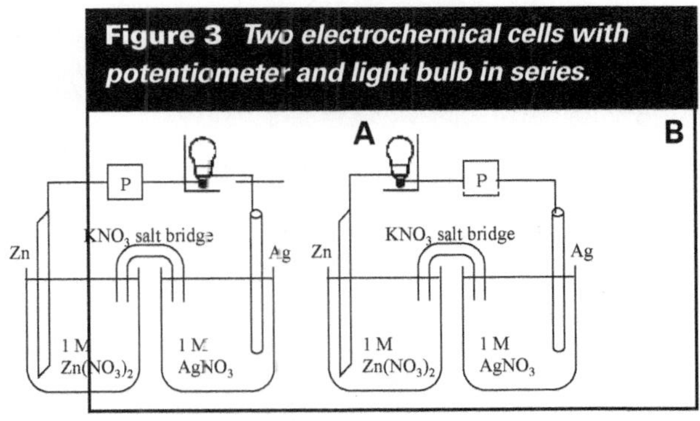

15. The light bulb does not illuminate in the other electrochemical cell either. Is this observation consistent with your explanation in the previous question? Explain your reasoning.

16. A potentiometer is used to read the voltage difference between two half-cells. What does "potential" refer to with a potentiometer (potential meter)?

17. The electrochemistry apparatuses A-F are grouped together in Table 1 at the end of the activity.

- Verify your answers to Q2 and Q4 by comparing to Table 1 entries for apparatuses A-D.
- Identify whether E and F are in the standard state and add your answers to Table 1.
- Using Q11, determine if E and F have complete circuits and add your answers to Table 1.

18. Consider again the collection of electrochemical apparatuses in Figure 2. For which of these cells do electrons flow freely (appreciable current) from the anode to cathode? Add your answers to the third row of Table 1.

Questions 19 through 23 refer to electrochemical cell A in Figure 2. Recall that the spontaneous reaction for this cell is the forward reaction: $Zn(s) + 2Ag^{+}(aq) \rightleftharpoons 2Ag(s) + Zn^{2+}(aq)$

19. What happens to the zinc metal after the spontaneous forward reaction has proceeded to a significant extent? Defend your choice.

(a) The Zn electrode increases in mass.

(b) The Zn electrode maintains its original mass.

(c) The Zn electrode has an overall smaller mass.

20. Consider the silver half-cell. Which of the following statements is true after the spontaneous forward reaction has proceeded to a significant extent?

(a) The Ag electrode decreases in mass.

(b) The $[Ag^{+}]$ decreases.

(c) The Ag electrode is unchanged as this reaction proceeds.

21. Based on your answers to Q19 and Q20, how does the concentration of cations (i.e., Ag^+ and Zn^{2+}) in solution change after the spontaneous forward reaction has proceeded to a significant extent?

The number of Ag^+ in the right half-cell:	increases	stays the same	decreases
The number of Zn^{2+} in the left half-cell:	increases	stays the same	decreases

22. Consider the graph below that shows the concentration of Zn^{2+} in cell A as a function of time.

(a) Why does one of the lines go up, while the other line goes down?

(b) Are the rates of change for the two lines different? Provide justification for you answer.

Figure 4 ***A graph of the concentration of each ion over time.*** *Disclaimer: The data shown here are representative not real data points.*

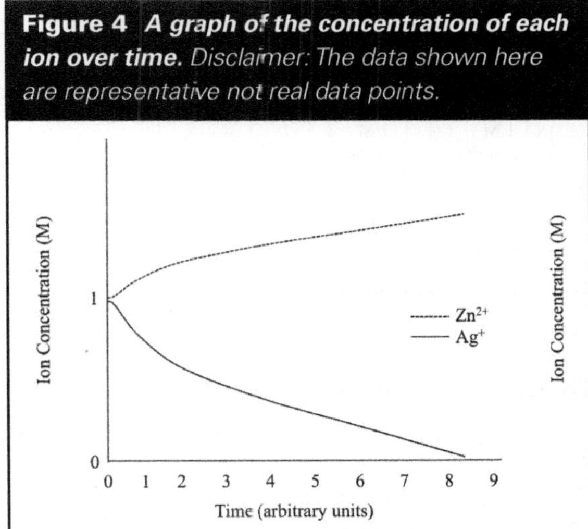

23. Recall the conditions for standard state defined on page 1. At which point(s) in time is electrochemical cell A in the standard state?

24. Electrochemical Cell F from Figure 2 appears below. Sketch the concentration profile for this cell.

Figure 2F ***Electrochemical apparatus with 1M solutions and a potentiometer.***

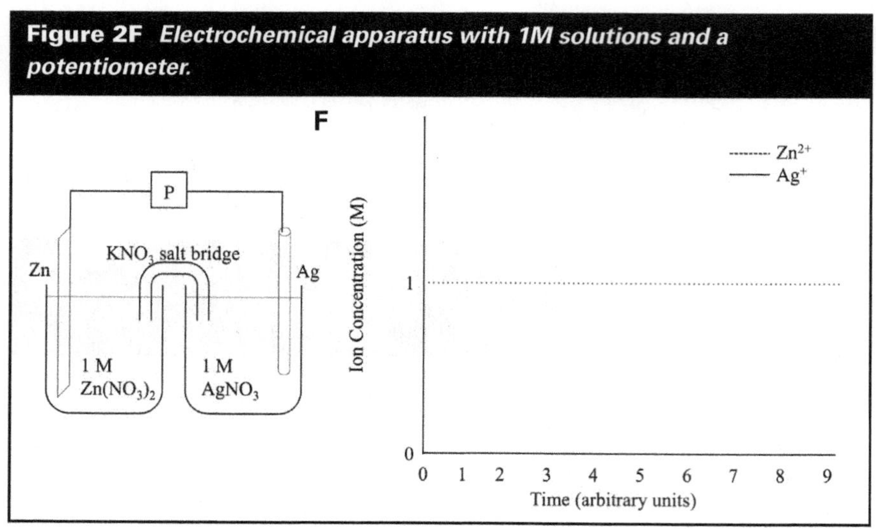

25. Consider all the electrochemical apparatuses in Figure 2. Which of these have concentrations that DO NOT change over time. Add your entries to the fourth row of Table 1.

26. How do you know if an electrochemical cell starts at 1 M and remains at 1 M?

Section 3: Calculating cell potentials

Consider this...

The potential difference between two electrodes in an electrochemical cell, E_{cell}, is determined by the Nernst equation:

$$E_{cell} = E^{\circ}_{cell} - \frac{RT}{nF} \ln Q = E^{\circ}cell - \frac{0.05916V}{n} \log Q$$

where E°_{cell} is the cell potential measured under standard conditions, n is the number of moles of electrons transferred in the electrochemical reaction, and Q is the reaction quotient for the forward reaction as written. Recall that the Nernst equation is frequently expressed in terms of base 10 logarithms for a cell at 25°C, in which case the ln term becomes the log term.

Key Questions

27. Write the Nernst equation expression for electrochemical cell F from Figure 2 (and shown here), given that the forward reaction below is spontaneous for this electrochemical cell.

$Zn(s) + 2Ag^{+}(aq) \rightleftharpoons 2Ag(s) + Zn^{2+}(aq)$

Figure 2F *Electrochemical apparatus with 1M solutions and a potentiometer.*

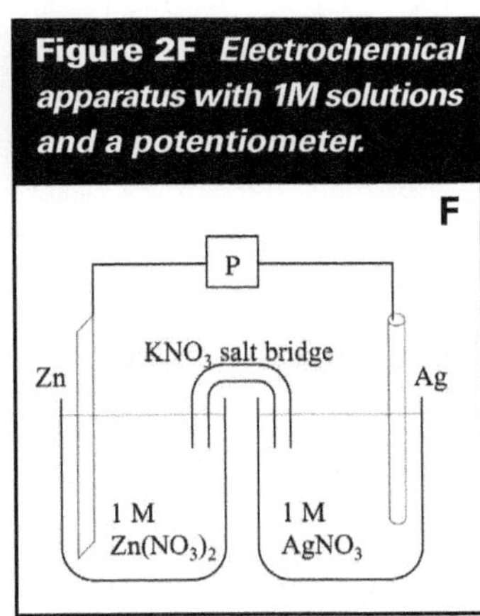

28. Look up the standard reduction potentials for the two half-reactions in electrochemical cell F in the table of standard reduction potentials provided in Q50 in the Applications section of this activity. Calculate E°_{cell} for the electrochemical reaction in Q27 under the conditions of cell F?

29. Calculate E_{cell} for electrochemical cell F using the Nernst equation expression written in Q27 and the $E°_{cell}$ value from Q28.

30. The Nernst equation cannot be used to calculate Ecell for the other five electrochemical apparatuses in Figure 2. For each electrochemical apparatus A-E specify the reason that Nernst equation can't be used.

A:

B:

C:

D:

E:

31. Are there any points in time that the Nernst equation can be used for any of the electrochemical apparatuses A-E? If so, indicate when it can be used.

Consider this...

The following electrochemical cells contain a Ag electrode immersed in $AgNO_3$ solution and a Pt electrode in a solution containing both Fe^{2+} and Fe^{3+}. We will consider each of these four cells separately.

Figure 5 ***Electrochemical apparatuses with the same anode and cathode but with varying concentrations of Fe^{2+}, Fe^{3+} and Ag^{+}.***

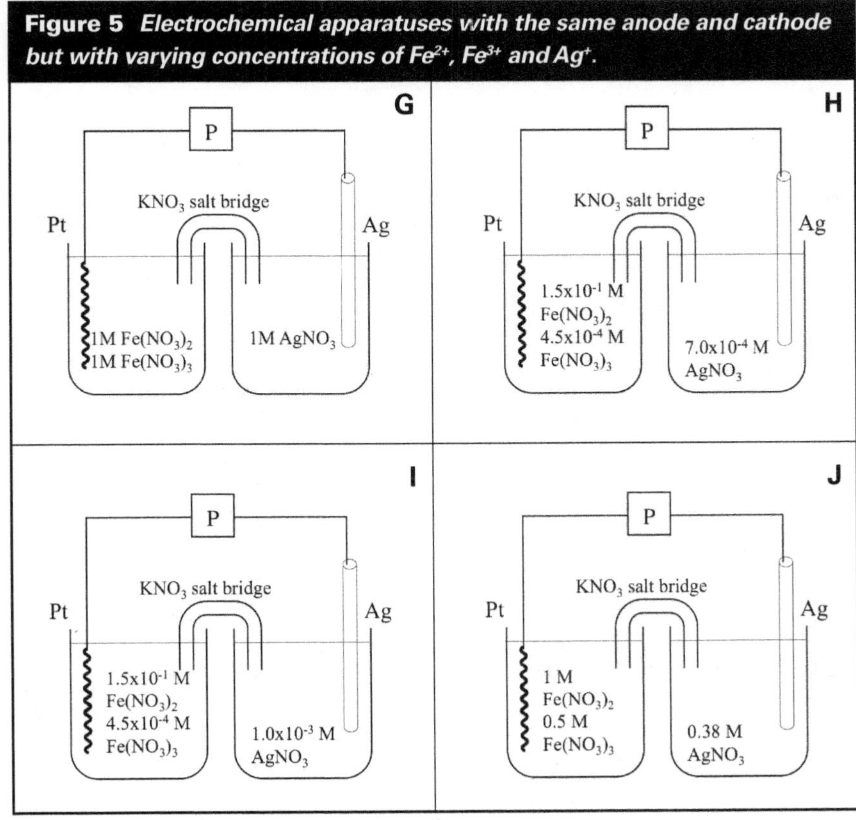

32. The reaction for electrochemical cell G is $Ag^{+} + Fe^{2+} \rightleftharpoons Ag(s) + Fe^{3+}$, which is spontaneous in the forward direction. Does E^0_{cell} indicate if this reaction at standard state is spontaneous? Use the Nernst equation to determine E_{cell} and for cell G. Add these values to Table 2 at the end of the activity.

33. Divide the remaining three cells H-J among group members. Use the Nernst equation to determine E_{cell} and $E°_{cell}$ for cells H-J. Add these values to Table 2.

34. Look at the E_{cell} and $E°_{cell}$ values in Table 2 for cells G-J.

a) Are all $E°_{cell}$ values equal? Why?

b) Is E_{cell} always greater than or equal to $E°_{cell}$?

35. Using the Nernst equation, specify when

a) $E_{cell} = E°_{cell}$

b) $E_{cell} < 0$

c) $E_{cell} > 0$

36. Using your prior knowledge of thermodynamics, what happens to a reaction when $E_{cell} < 0$ (i.e., $\Delta G > 0$)? Use "spontaneous" in your answer.

37. For cells H-J in Figure 5 decide which of the three statements about the reaction applies to each case. Mark your answers in Table 2.

A. The reaction proceeds in the forward direction: $Ag^+ + Fe^{2+} \rightleftharpoons Ag(s) + Fe^{3+}$

B. The reaction proceeds in the reverse direction: $Ag(s) + Fe^{3+} \rightleftharpoons Ag^+ + Fe^{2+}$

C. The reaction proceeds in neither direction because the reaction is at equilibrium

38. Support or refute this statement:

E°_{cell} *can be always be used to indicate the spontaneity of an electrochemical cell.*

Self-Assessment Questions

39. List one misconception that you had about electrochemical cells before doing this activity that has now been addressed.

40. How do you judge whether an electrochemical cell is in the standard state?

Applications

Note: In some instances, you will have to look up the standard reduction potential in the textbook to be able to complete the problem.

41. The electrochemical cells shown here have a zinc anode and a platinum electrode instead of silver in the other cells in this activity. The Zn electrode is immersed in a solution containing 1 M $Zn(NO_3)_2$, while the cathode is composed of a platinum electrode immersed in 1 M HNO_3 with H_2 (g) bubbled into the solution through a tube. The two metal electrodes are connected by a wire with a potentiometer (K) or light bulb (L) in the line. The KNO_3 salt bridge completes the electrical circuit between the separated half-cells.

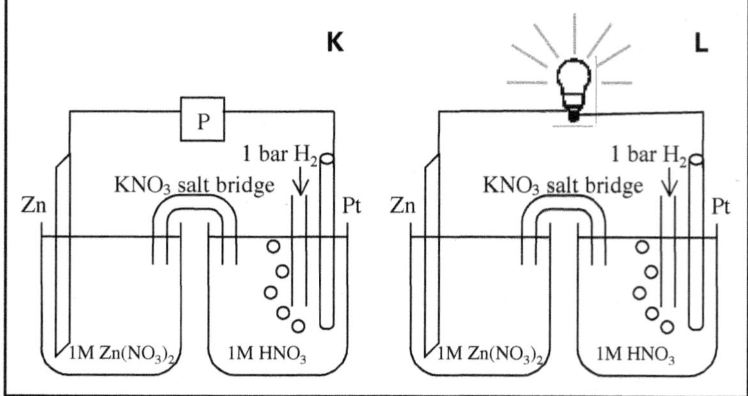

(a) Draw the path of H^+, Zn^{2+} and e- movement/flow on each of the figures above, as appropriate.

(b) Is Apparatus K in the standard state at any point after the circuit is complete? Is Apparatus L? Specify the time(s) at which each apparatus is the standard state.

(c) Calculate E_{cell} for apparatus K.

(d) Why can't E_{cell} be calculated for apparatus L?

42. Given this reaction:

$Zn(s) + Cu^{2+}(aq) \rightleftharpoons Zn^{2+}(aq) + Cu(s)$

(a) Sketch an electrochemical cell (with complete circuit) in the standard state.

(b) Calculate E_{cell} for this cell.

(c) As you have drawn the cell, will it remain in standard state after it is initially connected?

43. Calculate E_{cell} for this reaction if the solution concentrations are

$[Cr^{2+}] = 2.4 \times 10^{-2}$ M and $[Fe^{2+}] = 1.8 \times 10^{-3}$ M.

$Cr(s) + Fe^{2+}(aq) \rightleftharpoons Cr^{2+}(aq) + Fe(s)$

44. Consider electrochemical apparatus M at the right.

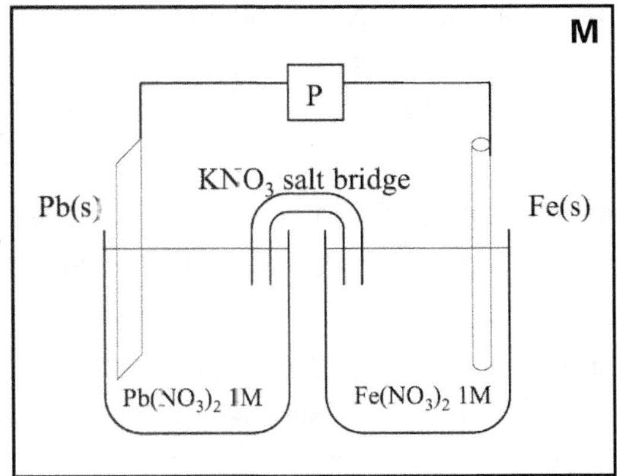

(a) What is the overall spontaneous reaction for electrochemical apparatus M? Calculate E_{cell}.

(b) How would E_{cell} change if the potentiometer were removed and replaced with a wire connecting the Pb and Fe electrodes?

(c) Suppose the cell with the wire connection is allowed to run until the potential difference between the two electrodes was 0.28 V. At this point in the discharge of the electrochemical cell what are [Fe^{2+}] and [Pb^{2+}]?

45. Compare and contrast an electrochemical cell that has a light bulb connecting the two electrodes (like Electrochemical Cell A in Figure 2) and one with a potentiometer connecting the two electrodes (like Electrochemical Cell F in Figure 2).

46. The graph shows a concentration vs. time profile for three different half-cells with the reaction $M^{n+} + n\ e^- \rightleftharpoons M(s)$. The concentration refers to the M^{n+} ion in solution as the cell is set up as part of an electrochemical cell and allowed to run.

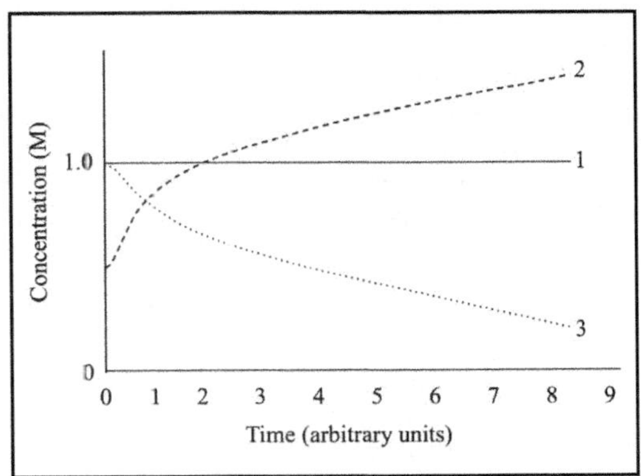

(a) For each cell 1-3, specify whether the cell is ever in the standard state. If it is, then specify at which times during the experiment it is in the standard state.

(b) Imagine the half-cells 1-3 are connected to another half-cell with a completed circuit. For which of the half-cells 1-3 does current flow? Explain how you decided.

(c) Since the half-reaction corresponding to each of these concentration profiles is $M^{n+} + n\ e^- \rightleftharpoons M(s)$, in which cases 1-3 does the reaction occur as written (i.e., reduction of the metal ion to solid metal)?

47. You decide to construct an electrochemical cell to power your fan. Your fan requires a voltage of 1.50 V in order to operate. You have the following three half-cells to combine to form a complete cell generating a voltage sufficient to power your fan. In each half-cell the metal ion concentration is 1.0 M.

Reaction	**E° (Volts)**
$Ag^+ + e^- \rightleftharpoons Ag(s)$	0.799
$Pb^{2+} + 2e^- \rightleftharpoons Pb(s)$	-0.126
$Zn^{2+} + 2e^- \rightleftharpoons Zn(s)$	-0.763

(a) Which pair of half-cells do you choose?

(b) If the clock draws 1.0 mA of current, how long will it run until the voltage drops below 1.50 V?

(c) Does it matter which electrode is connected to the positive terminal of your fan? Why or why not?

48. Suppose the wire connecting the Zn and Ag electrodes attached to the light bulb is removed from Electrochemical Cell A and the leads are reversed so the end originally attached to the Zn electrode is attached to the Ag electrode and vice versa. Will the light bulb produce light? Explain your reasoning.

49. If the potentiometer (which measures voltage) in Electrochemical Cell F is connected in the opposite direction (i.e., the leads are switched) how will the reading on the meter change from the value it had originally?

50. The Table below lists Standard Reduction Potentials for several half-reactions. For what concentrations of solution species are these E° values valid?

Standard Reduction Potentials for Selected Half-Reactions	
Reaction	**E° (Volts)**
$MnO_{4^-} + 8\ H^+ + 5\ e^- \rightleftharpoons Mn^{2+} + 4\ H_2O$	1.507
$Ag^+ + e^- \rightleftharpoons Ag(s)$	0.799
$Fe^{3+} + e^- \rightleftharpoons Fe^{2+}$	0.771
$Cu^{2+} + 2e^- \rightleftharpoons Cu(s)$	0.339
$2H^+ + 2e^- \rightleftharpoons H2(g)$	0.000
$Pb^{2+} + 2e^- \rightleftharpoons Pb(s)$	−0.126
$Cd^{2+} + 2e^- \rightleftharpoons Cd(s)$	−0.402
$Fe^{2+} + 2e^- \rightleftharpoons Fe(s)$	−0.44
$Zn^{2+} + 2e^- \rightleftharpoons Zn(s)$	−0.762
$Cr^{2+} + 2e^- \rightleftharpoons Cr(s)$	−0.89

51. Using the axes below, sketch how the potential difference between the Cu and Zn electrodes will change as a function of time for Electrochemical Cell N and for Electrochemical Cell O, also shown below.

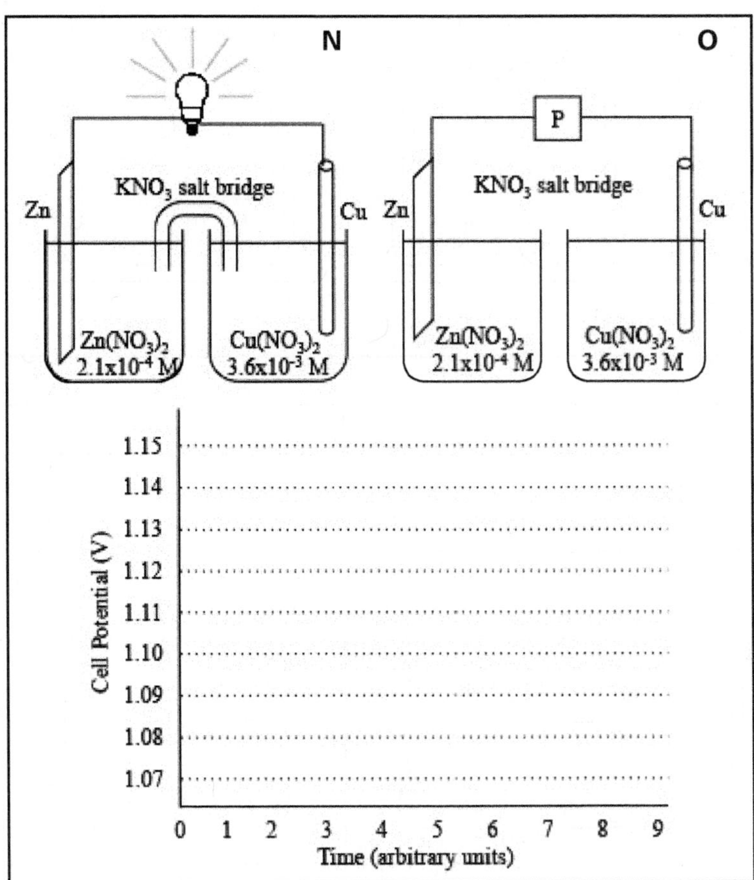

Table 1

	A	B	C
Electrochemical Apparatus	Zn; KNO_3 salt bridge; Ag; 1 M $Zn(NO_3)_2$; 1 M $AgNO_3$	Zn; KNO_3 salt bridge; Ag; $1.3x10^{-3}$ M $Zn(NO_3)_2$; $2.5x10^{-4}$ M $AgNO_3$	Zn; KNO_3 salt bridge; Ag; 1 M $Zn(NO_3)_2$; 1 M $AgNO_3$
Standard State Concentrations?	Yes	No	Yes
Complete Circuit?	Yes	Yes	No
Electrical Current?			
Concentrations Fixed?			
	D	**E**	**F**
Electrochemical Apparatus	Zn; Ag; $1.3x10^{-3}$ M $Zn(NO_3)_2$; $2.5x10^{-4}$ M $AgNO_3$	Zn; KNO_3 salt bridge; Ag; 1 M $Zn(NO_3)_2$; 1 M $AgNO_3$	P; Zn; KNO_3 salt bridge; Ag; 1 M $Zn(NO_3)_2$; 1 M $AgNO_3$
Standard State Concentrations?	No		
Complete Circuit?	No		
Electrical Current?			
Concentrations Fixed?			

Table 2

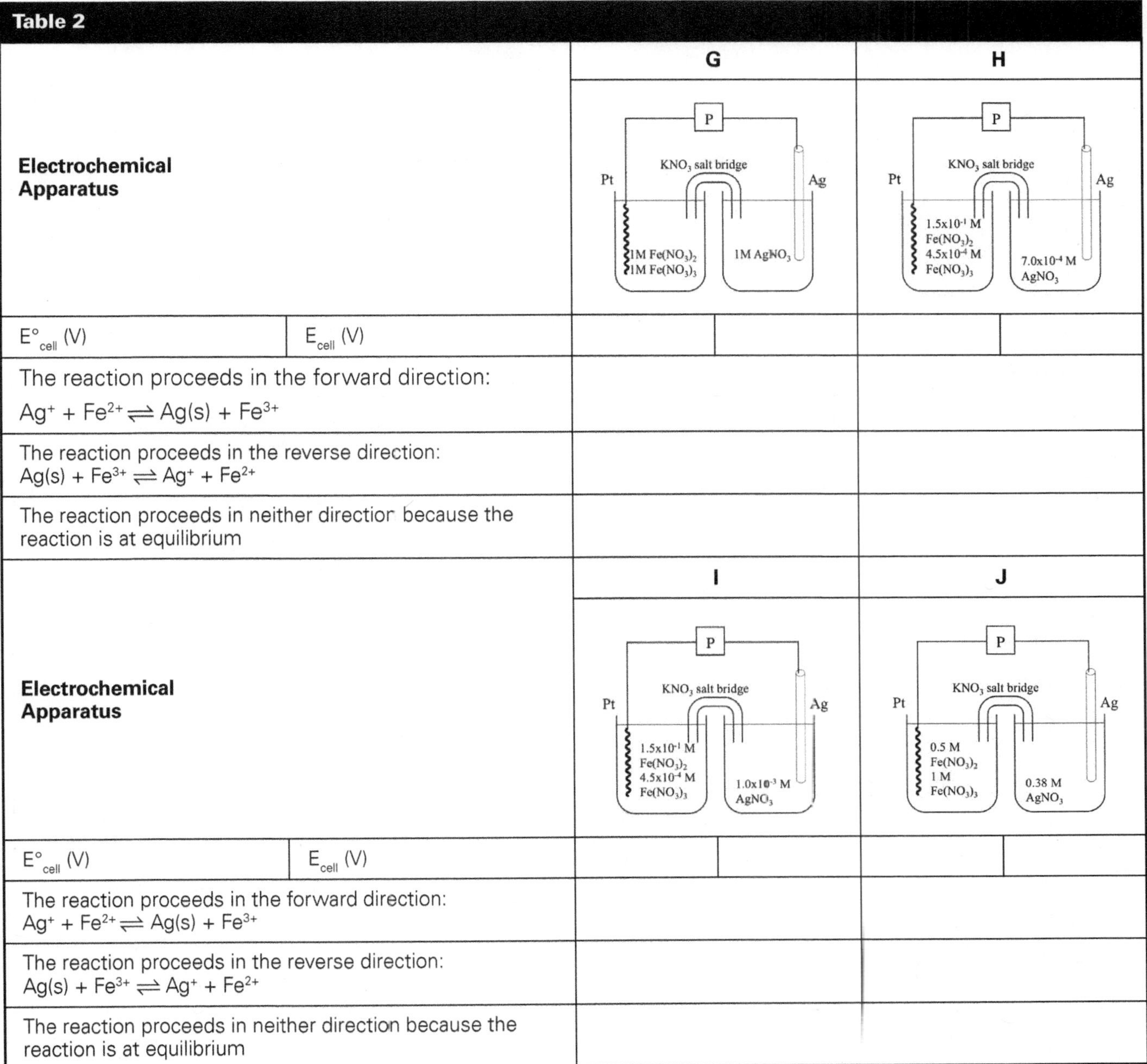

		G		H	
Electrochemical Apparatus					
E°_{cell} (V)	E_{cell} (V)				
The reaction proceeds in the forward direction: $Ag^+ + Fe^{2+} \rightleftharpoons Ag(s) + Fe^{3+}$					
The reaction proceeds in the reverse direction: $Ag(s) + Fe^{3+} \rightleftharpoons Ag^+ + Fe^{2+}$					
The reaction proceeds in neither direction because the reaction is at equilibrium					
		I		**J**	
Electrochemical Apparatus					
E°_{cell} (V)	E_{cell} (V)				
The reaction proceeds in the forward direction: $Ag^+ + Fe^{2+} \rightleftharpoons Ag(s) + Fe^{3+}$					
The reaction proceeds in the reverse direction: $Ag(s) + Fe^{3+} \rightleftharpoons Ag^+ + Fe^{2+}$					
The reaction proceeds in neither direction because the reaction is at equilibrium					

The Beer-Lambert Law

Learning Objectives

Students should be able to:

Content

- Identify each term in the Beer-Lambert law and explain its effect on absorbance.
- Explain the relationship between transmittance and absorbance.
- Use Beer's Law in quantitative measurements.

Process

- Prepare and interpret graphs (information processing).
- Develop mathematical expressions to describe data (problem solving).

Prior knowledge

- Calibration curves and linear equations.

Further Reading

- D.C. Harris, *Quantitative Chemical Analysis,* 7th Edition, 2007 W.H. Freeman: USA, Section 18-2, pp. 380.

Authors

Caryl Fish, David Langhus

Consider this...

Suppose that we have a solution that absorbs light.

Suppose further, that we propose to analyze the solution using an instrument organized as shown below:

Figure 1 ***Simple spectrometer***

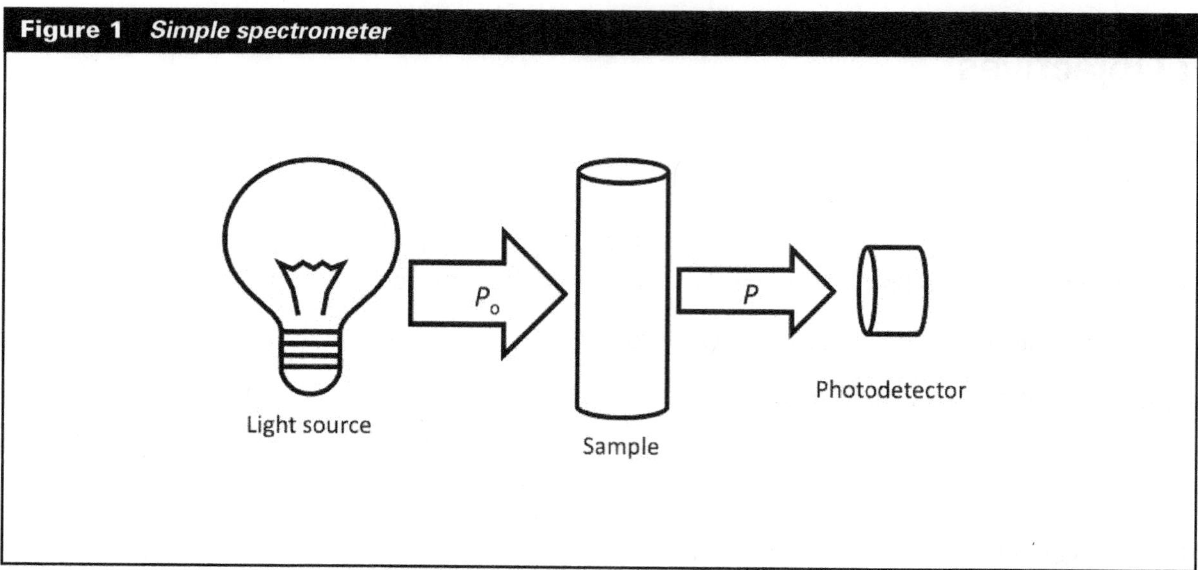

The light source might be an ordinary light bulb, a flashlight, or the sun. The photodetector could be a simple solar cell like one might buy at Radio Shack. The light shining on the sample has power P_0 and the light that manages to get through the sample has power P. P_0 and P may be expressed in Lamberts or lumens or ergs/sec/cm^2 or any other units, as long as the same units are used for both.

Transmittance is defined as $T = P/P_0$.

Percent transmittance, or $\%T$, is simply 100 T.

Key Questions

1. If some of the light is absorbed by the sample....

Is P greater than, less than, or equal to P_0? Justify your answer to your group members.

Should be less than P_0

Is T greater than, less than, or equal to 1? Justify your answer to your group members.

Less than 1.

2. Absorbance indicates how much light is absorbed by a sample. If sample A (below) absorbs more light than sample B, which would have the higher transmittance? Which would have the higher absorbance? Explain your reasoning.

Figure 2 ***Two absorbing samples***

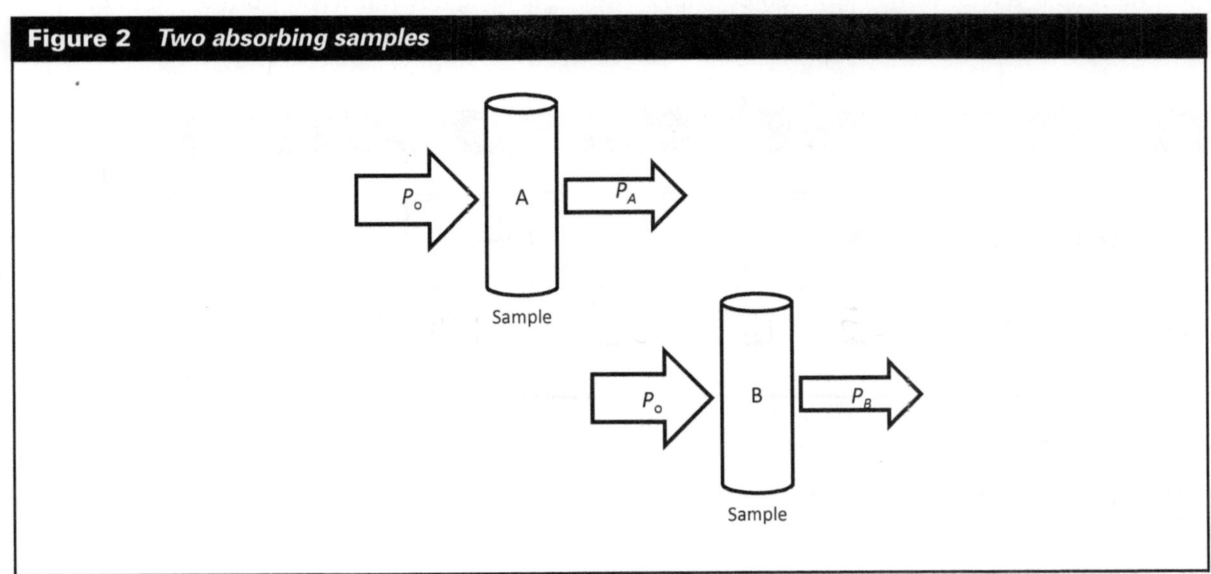

B should have more transmittance. Less light is absorbed, so more light gets through the sample to the detector. A has the higher absorbance because "Sample A absorbs more light."

3. Discuss with your group members what happens to transmittance if absorbance increases or decreases. In 2-3 complete sentences explain why transmittance and absorption are related in this way.

If absorbance increases / decreases, transmittance decreases / increases.

Absorption prevents light from moving completely through the sample. Therefore, a sample that absorbs less light will allow for greater transmittance of light.

Consider this...

Suppose that we have multiple identical samples of a light-absorbing solution (in containers that do not absorb light) and we set them up in a line, with a light source with power P_0 shining down the line from one end as shown in Figure 3.

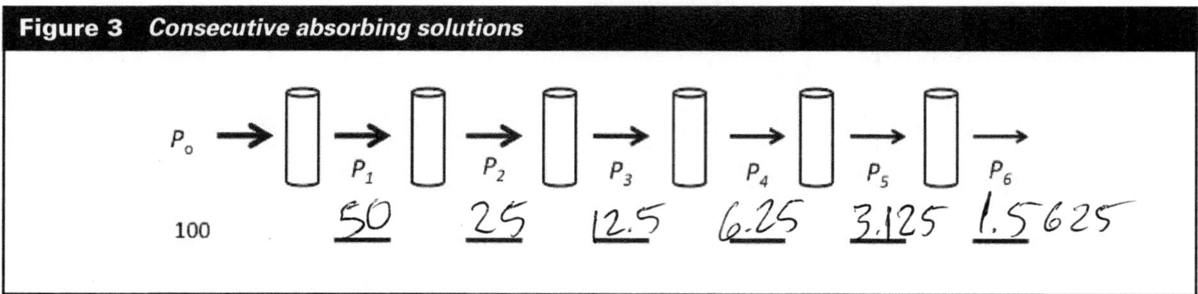

Figure 3 ***Consecutive absorbing solutions***

Key Questions

4. If 50% of the light is transmitted through each sample and the initial power, P_0, is 100 arbitrary units, calculate values for P_1 through P_6 and fill in the blanks in the figure with these values. Compare your values with your group members.

5. Using the equation for transmittance, $T = P/P_0$, each group member should calculate the total transmittance value for two different n values where n is the number of solutions through which light passes. Share your answers and graph the T values *vs. n*:

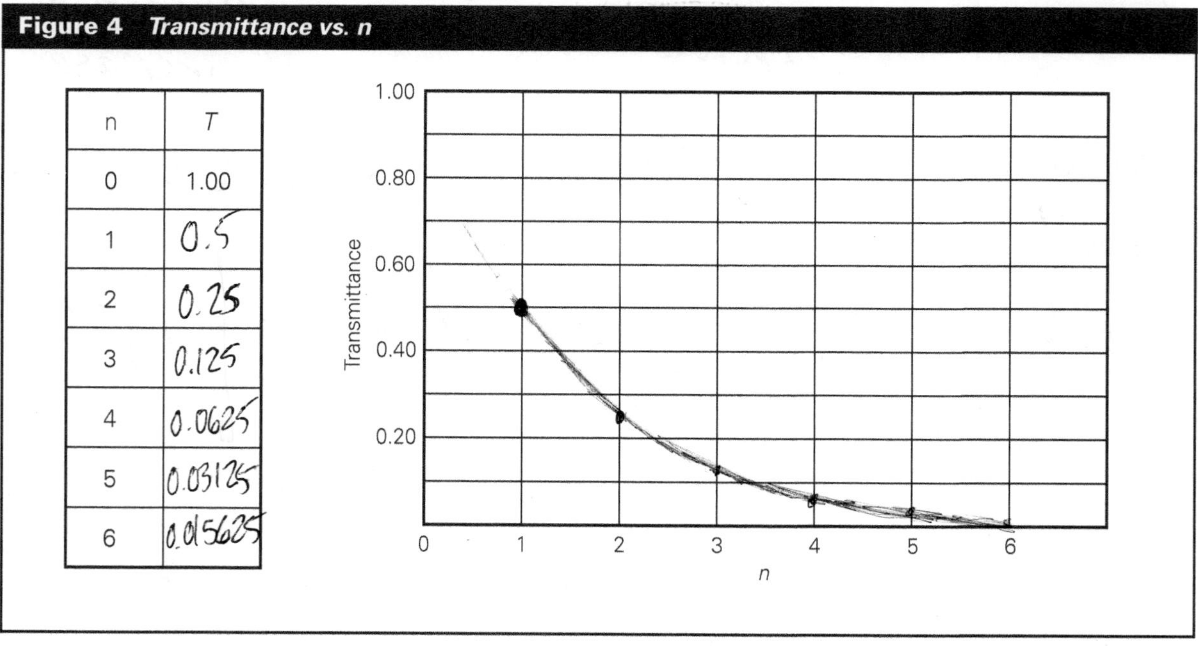

n	T
0	1.00
1	0.5
2	0.25
3	0.125
4	0.0625
5	0.03125
6	0.015625

Figure 4 ***Transmittance vs. n***

6. Does the transmittance increase or decrease as *n* increases? Is there a linear relationship between the transmittance and the number of samples through which the light passes?

Transmittance decreases as n increases.
It is not a linear relationship

7. Repeat the exercise above but this time plot ($-\log_{10}T$) *vs. n* (number of solutions through which light passes).

Figure 5 *($-\log_{10}T$) vs. n*

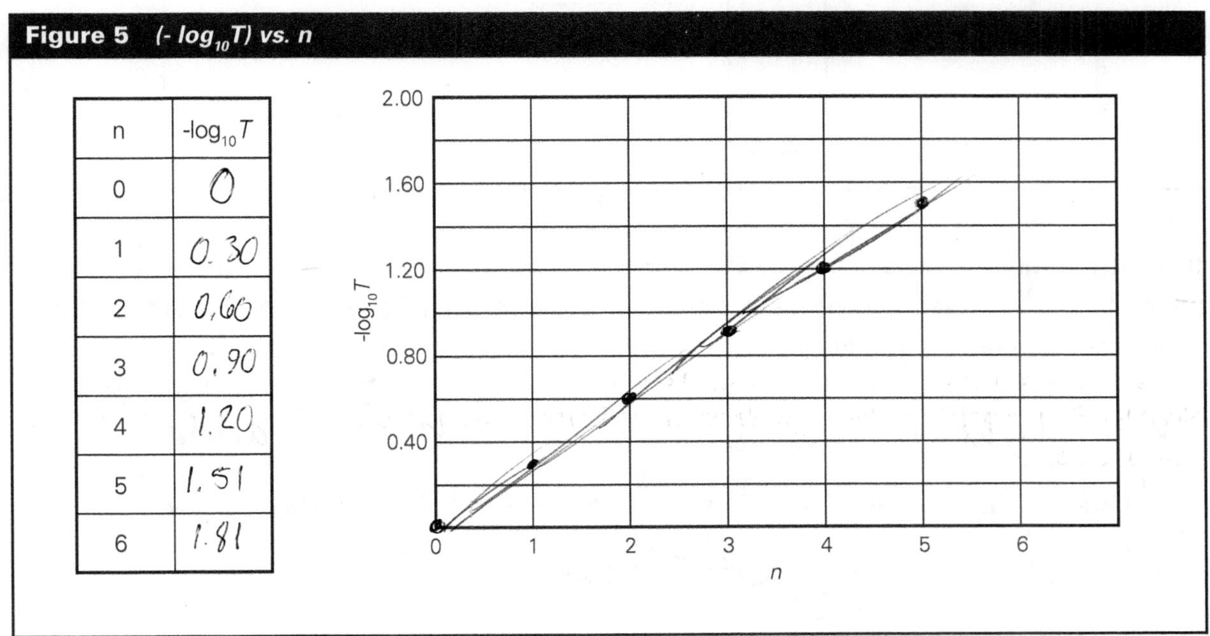

n	$-\log_{10}T$
0	0
1	0.30
2	0.60
3	0.90
4	1.20
5	1.51
6	1.81

Does ($-\log_{10}T$) appear to vary linearly with *n*?

it's linear

8. Using the standard form for a linear equation ($y = mx + b$, where m is the slope and b is the intercept), write an equation for the relationship shown on the graph in Figure 5.

$$-\log_{10}T = 0.30n + 0$$

What value should be expected for the intercept?

zero

Now it is convenient to describe $-\log_{10} T$ as the absorbance *(A)* such that $A = -\log_{10} T$.

Figure 6 ***A sample can be considered a series of identical thin slices***

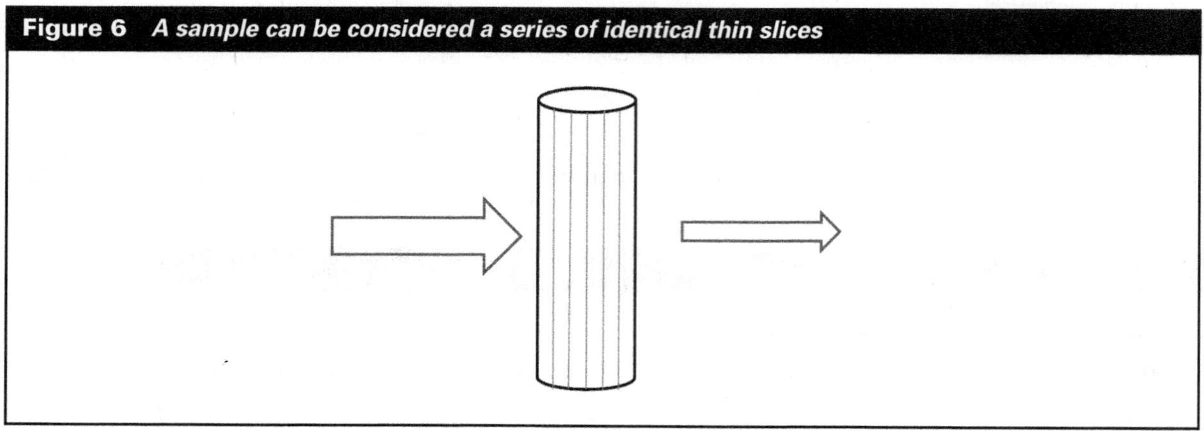

9. Suppose that each individual sample in the line of samples in Figure 3 is a slice of a mega-sample. In other words, instead of six successive solutions there is one large container with 6 slices of solution put together (Figure 6). Look back at your data in Figure 5, but now think of n as the number of slices in the sample. In words, describe the relationship between distance the light travels through the sample and $-\log_{10} T$ (or *A*).

More light is absorbed with each slice added.
Each slice cuts the transmittance by half and doubles the absorbance.

10. Using b to represent the total distance the light travels—the pathlength—modify your mathematical expression in Q8 to express the relationship between absorbance *(A)* and pathlength *(b)*.

$A = 0.30 b$

Consider this...

Below are three samples with different amounts of absorbing compound which are represented by ✸.

The distance the light travels through each sample (pathlength) and the power of the incoming light (P_0) are the same for each sample.

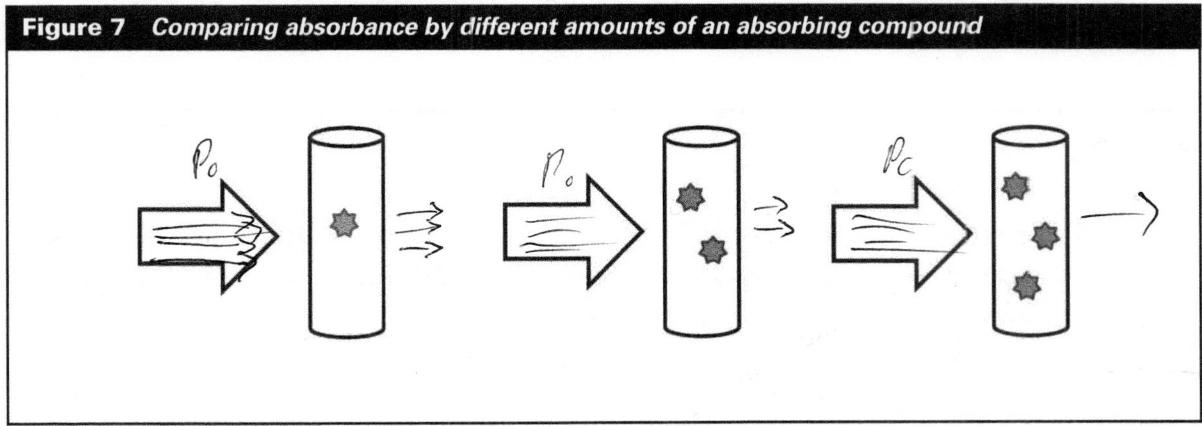

Figure 7 *Comparing absorbance by different amounts of an absorbing compound*

11. Compare the concentrations of the absorbing species in the three samples. What assumption are you making about the volume of the samples?

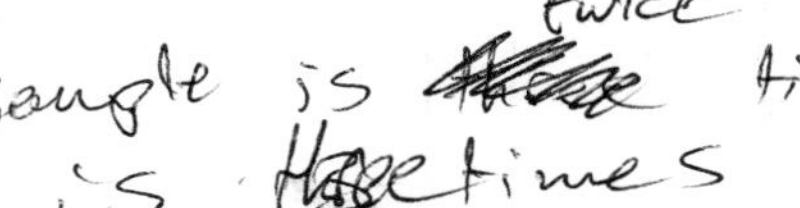 

The second sample is twice times and the third

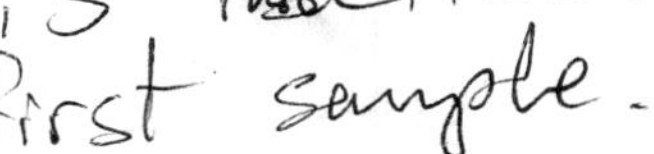 

sample is three times the concentration of the first sample.

Assume volume is the same, or it doesn't matter

12. Discuss with your group the relationship between concentration (c) and absorbance. Draw arrows in Figure 7 to represent the transmitted light through each sample. Modify your mathematical expression in question 10 to include both pathlength and concentration.

concentration increases → absorbance increases

$A = 0.30bc$

The absorbance, A, of a solution is related to the distance through which light travels in a sample and the concentration of the sample by means of the Beer-Lambert Law or simply Beer's Law:

$A = abc$

a = absorptivity, a characteristic of a substance that determines extent of absorption

b = pathlength, distance light travels through the sample.

c = concentration of absorbing species in the sample

Consider this...

A spectrophotometer is an instrument that can measure the absorbance of a sample at different wavelengths of light. A graph of the absorbance of a sample as a function of wavelength obtained from such an instrument is called a spectrum and is shown in Figure 8. The spectrum shown below is for a 100 ppm solution of Compound X using a 1.00 cm wide sample vial.

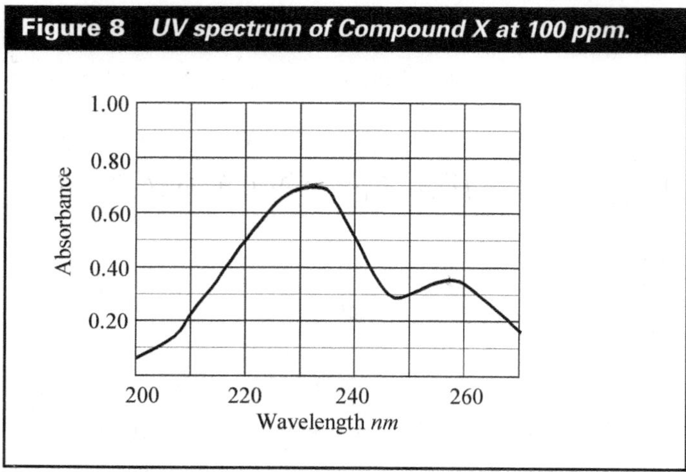

Figure 8 ***UV spectrum of Compound X at 100 ppm.***

Key Questions

13. From the description above this spectrum, what terms in Beer's Law have been held constant to obtain the spectrum in Figure 8?

pathlength, concentration

14. What is the absorbance of Compound X under these conditions at 233 nm? At 257 nm?

0.70 ; 0.37

15. Discuss with your group how the data collected in Figure 8 illustrates Beer's Law. What term in Beer's Law must be wavelength-dependent?

Assuming path length and concentration are constant, Absorbance ~~is proportional to the~~ wavelength-dependent only.

16. Beer's Law assumes monochromatic light, but monochromatic light is not very practical to use for most analyses. Choosing a wavelength with little variation in absorbance over a short range of wavelengths can approximate monochromatic light and therefore Beer's Law will hold. What wavelengths from Figure 8 would best meet this criterion?

233nm ; 257nm

17. Determine with your group which wavelength in Figure 8 would allow you to measure the lowest concentration of Compound X and explain why this wavelength was chosen.

257nm gives lowest absorbance

18. Sketch the absorbance vs. concentration plot for Compound X at both 233 nm and 257 nm. How are they similar and how are they different?

- Both are linear.
- 233nm is steeper because it absorbs more light at higher concentrations.

233nm — abs / conc.

257nm — abs / conc.

19. What are two things you did not know before, but learned about Beer's Law during this activity?

- some significance of each absorption peak as being better at absorbing lower or higher concentrated samples

20. In what ways did this activity enhance your ability to write mathematical expressions to describe relationships?

explanation of context for some of the mathematical relationships

Applications

21. What is the value of absorbance that corresponds to 35.0%T? Spacing seemed off here.

22. At very low values of T (less than 5%), describe the absorbance values. What problems could arise in detecting the light passing through this sample?

23. Molar absorptivity (ε) is the absorptivity when concentration is in units of molarity and the pathlength is in centimeters. If a 1.53 x 10^{-5} M solution has an absorbance of 0.426 in a 1.00 cm cell when measured at a wavelength of 254 nm, what is the molar absorptivity of the compound?

24. Iron (II) is measured in abandoned mine drainage using a phenanthroline method (Standard Methods for the examination of water and wastewater, method 3500-Fe). In this method 1.00 mL of duplicate water samples are acidified, added to phenanthroline solution and diluted to 100.0 mL. The solution turns deep red and is measured at 510 nm. A series of standards from 0.50-2.00 μg/mL Iron(II) are prepared and their absorbance measured. The data for these solutions is given below. Using this data calculate the concentration of iron(II) in the original mine drainage samples.

Fe Conc (μg/mL)	Absorbance
0.00	0.002
0.50	0.045
1.00	0.096
1.50	0.134
2.00	0.196
Mine drainage	0.107 0.103

25. Fresh meat slowly changes from bright red to a brown color due to the oxidation of Oxy-myoglobin to met-myoglobin upon aging. The concentrations of both of these proteins can be measured in ground meat (after extraction, purification, and separation) using a spectrophotometer. The oxy-myoglobin is measured at 417 nm with a molar absorbtivity of 12800 $M^{-1}cm^{-1}$. The met-myoglobin is measured at 409 nm with a molar absorbtivity of 17900 $M^{-1}cm^{-1}$. In an extract of 10.0 g of ground meat (in 3.0 mL total solution) the absorbance of the oxy-myoglobin sample was found to be 0.769 using a 1 cm cell and the met-myoglobin was 0.346. Determine the moles of oxy-myoglobin and met-myoglobin per gram of ground meat. (Bylkas, S.A.; and Andersson, L.A. J. *Chem. Ed.* 1997, 74, 426-430.)

26. A solution suspected of containing copper is analyzed. To determine the standard curve, the absorbance of five standards ranging from 0.125 mg/L Cu to 1.25 mg/L Cu was measured and graphed against the concentration. The slope of the linear best-fit line was determined to be 0.6136 and the y-intercept was 0.0142. If three samples of the unknown solution have absorbance values of 0.438, 0.434, and 0.439 what is the concentration of the unknown?

27. A solution of Compound X is to be measured using Beer's Law. Based on Figure 8, what are two advantages of choosing a wavelength of 233 nm instead of 240 nm?

28. A spectrophotometer will often produce values of both Absorbance and % Transmittance. Which would be the easiest for you to use to determine the concentration of a compound in solution? Explain.

29. Suppose that in measuring a sample of iron using an absorption spectrophotometer, you find that the absorbance is too high for the instrument to measure accurately. Based on Beer's Law, what are two options you might employ to lower the absorbance value?

30. In an Atomic Absorption (AA) instrument a flame is the sample holder. In the configurations below would the short, wide flame or the tall, narrow flame give the highest absorbance values for the same concentration of sample? Explain why.

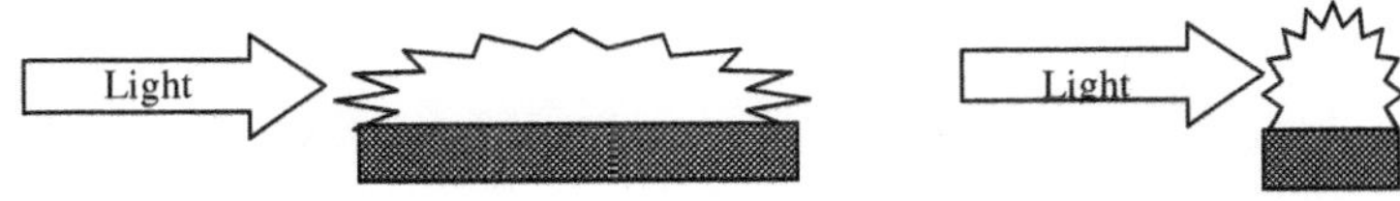

31. Beer's Law has some limitations. It works well for monochromatic radiation, dilute solutions, and when the absorbing compounds are not participating in a concentration-dependent equilibrium. Plot the following data using Excel and determine the highest concentration in which Beer's Law applies.

Conc (mg/L)	Absorbance
1.00	0.0334
5.00	0.1635
10.00	0.3284
15.00	0.4965
20.00	0.6026
30.00	0.7781
50.00	0.8652

Atomic and Molecular Absorption Processes

Learning Objectives

Students should be able to:

Content

- Relate the wavelength and energy of light absorbed to molecular and atomic transitions.
- Compare molecular and atomic absorption processes and explain differences in band width in the resulting spectra.

Process

- Compare and contrast spectra and absorption processes (critical thinking).

Prior knowledge

- Electromagnetic spectrum.
- Energy and Wavelength relationships.
- Molecular orbital terminology.

Further Reading

- Skoog, D.A., Holler, J.F., Crouch, S.R. *Principles of Instrumental Analysis,* Sixth Edition, Thomson Brooks/Cole, 2007, pp. 131-164 and 336-367.
- D.C. Harris, *Quantitative Chemical Analysis,* 7th Edition, 2007 W.H. Freeman: USA, Section 18-5, pp. 387-389.
- A.M. Bass, L.C. Glasgow, C. Miller, J.P. Jesson, and D.L. Filkin, "Temperature dependent absorption cross sections for formaldehyde (CH_2O): The effect of formaldehyde on stratospheric chlorine chemistry," Planet. Space Sci. 28, 1980, 675-679.

Authors

Caryl Fish, Ruth Riter

Consider this...

Table 1 *Electromagnetic Radiation*

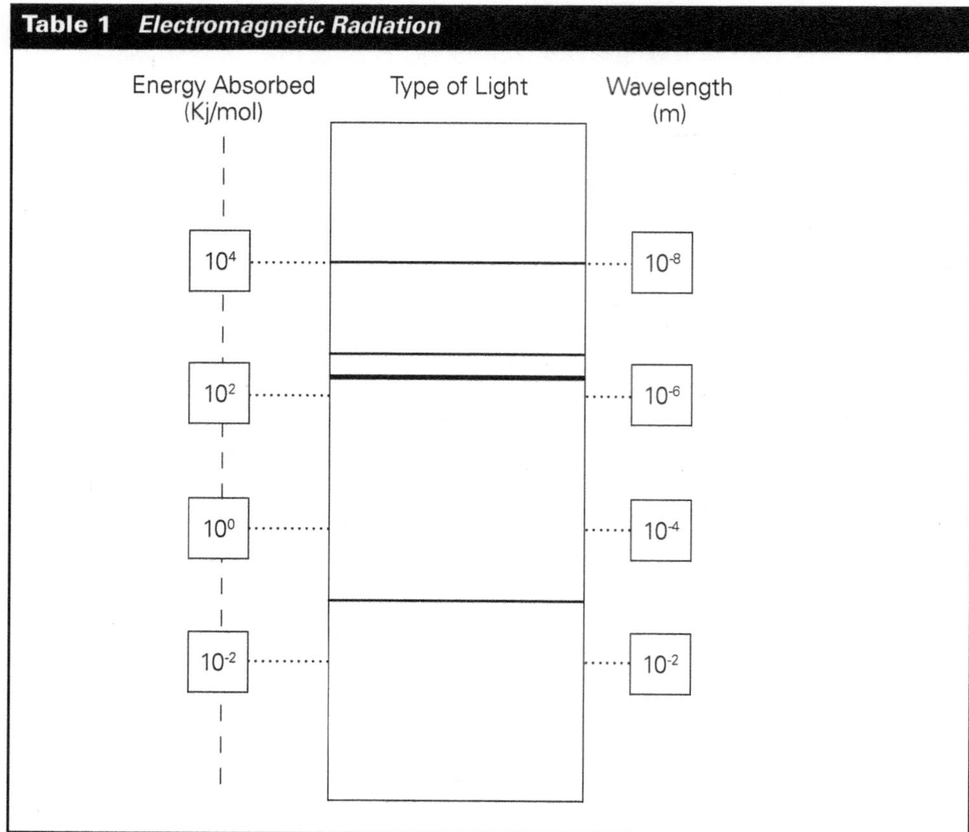

Key Questions

1. How are the energy and wavelength related in the table?

2. Recall the relationship between wavelength and frequency. Is the table consistent with E=hv? Explain your reasoning.

3. Add the types of electromagnetic radiation below to Table 1, above.

a. Infrared light (λ=7.80 x 10^{-7} – 1 x 10^{-3} m)

b. Ultraviolet light (λ=1 x 10^{-8} – 3.80 x 10^{-7} m)

c. X-rays (λ=10^{-11} to 10^{-8} m)

d. Microwave (λ=10^{-3} to 10^{-1} m)

e. Visible light (λ= 3.80 x 10^{-7} to 7.80 x 10^{-7} m)

Consider this…

Figure 1 *Atomic Absorption of Sodium*

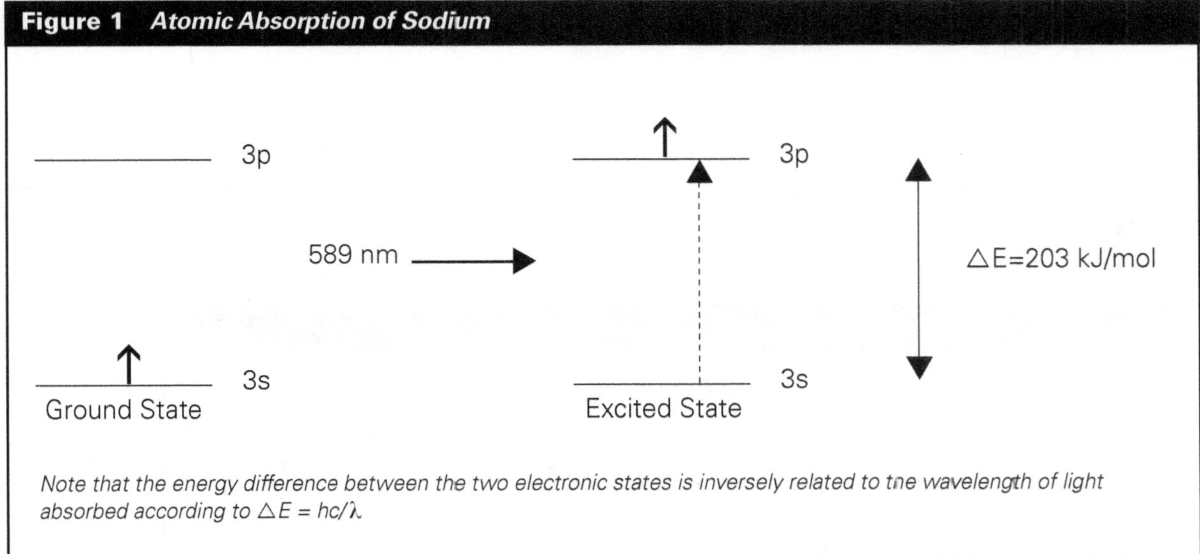

Note that the energy difference between the two electronic states is inversely related to the wavelength of light absorbed according to $\Delta E = hc/\lambda$.

Key Questions

4. In Figure 1, the outermost electron of sodium resides in which ground state atomic orbital?

5. The electron is in which atomic orbital after the atom absorbs a photon of light?

6. How much energy is absorbed in this transition?

7. Sodium absorbs light at both 589.0 and 589.6 nm due to differences in spin of the outermost electron. At this point assume equal absorbance at each wavelength. Using this information, draw a sketch that represents wavelength vs. absorbance for sodium.

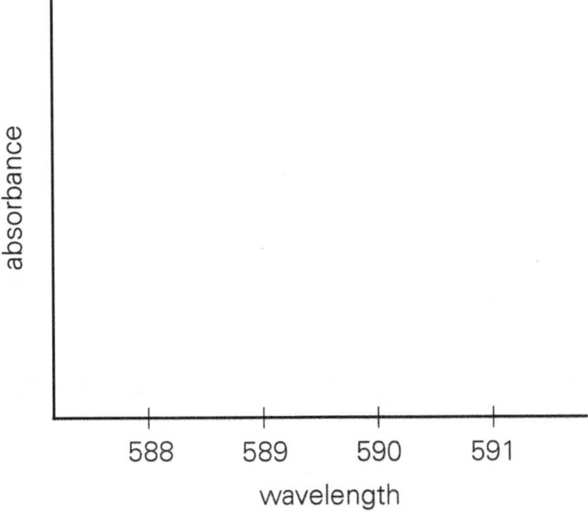

In atomic absorption the absorption bands are very narrow, and less than a nanometer wide.

8. After reading the above information, compare your sketch with your group members and adjust your sketch if necessary.

Consider this...

Figure 2 ***Electronic Transition for Formaldehyde***

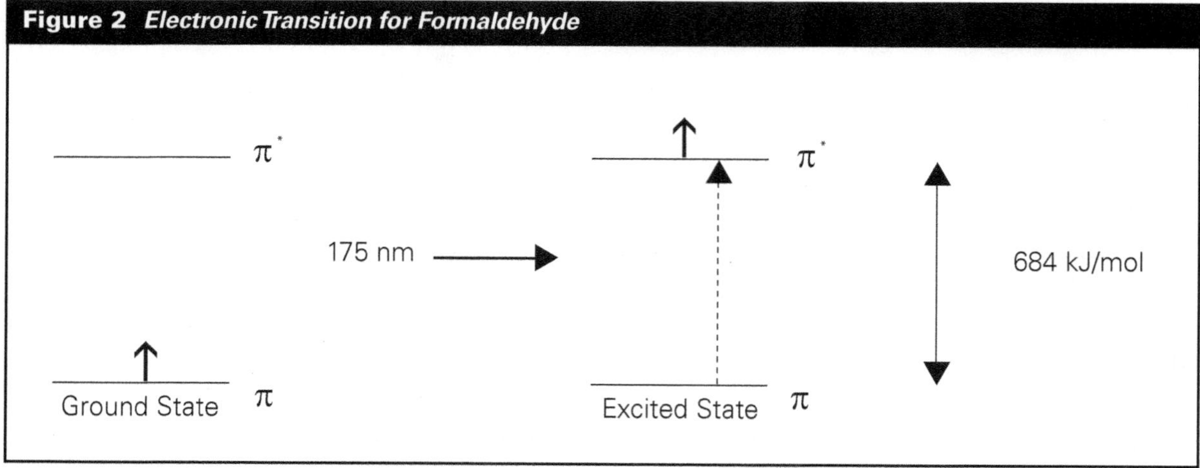

Key Questions

9. In Figure 2, what do the π and π* symbols represent?

10. In what ways is this diagram of molecular absorption different than the atomic absorption diagram?

How is it similar?

11. Based on the wavelength, the absorption of what type of electromagnetic radiation causes the electronic transition in formaldehyde?

Consider this...

Molecules also have vibrational and rotational energy states. Below are vibrational and rotational transitions for formaldehyde.

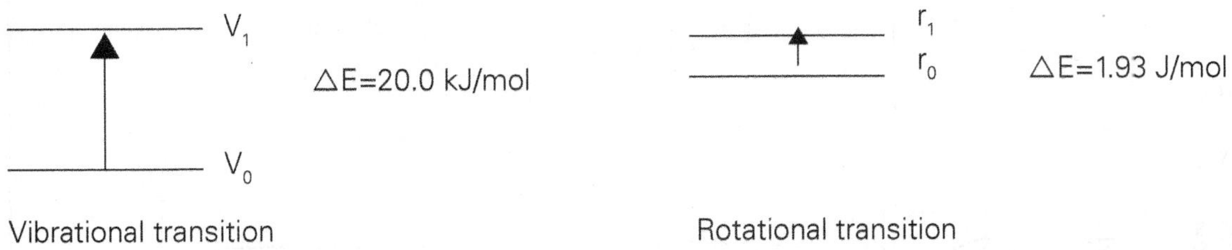

Vibrational transition

Rotational transition

Key Questions

12. Compare the energy required for the electronic, vibrational, and rotational transitions of formaldehyde. Which requires the most energy? Which the least?

13. The energy required for the vibrational transition is about (10, 100, 1000, or 10,000) times more than the energy required for the rotational transition.

The energy required for the electronic transition is about (10, 100, 1000 or 10,000) times more than the energy required for the vibrational transition.

14. Revisiting our original table, now add the molecular processes of electronic, vibrational, and rotational excitation to the appropriate boxes in the table below.

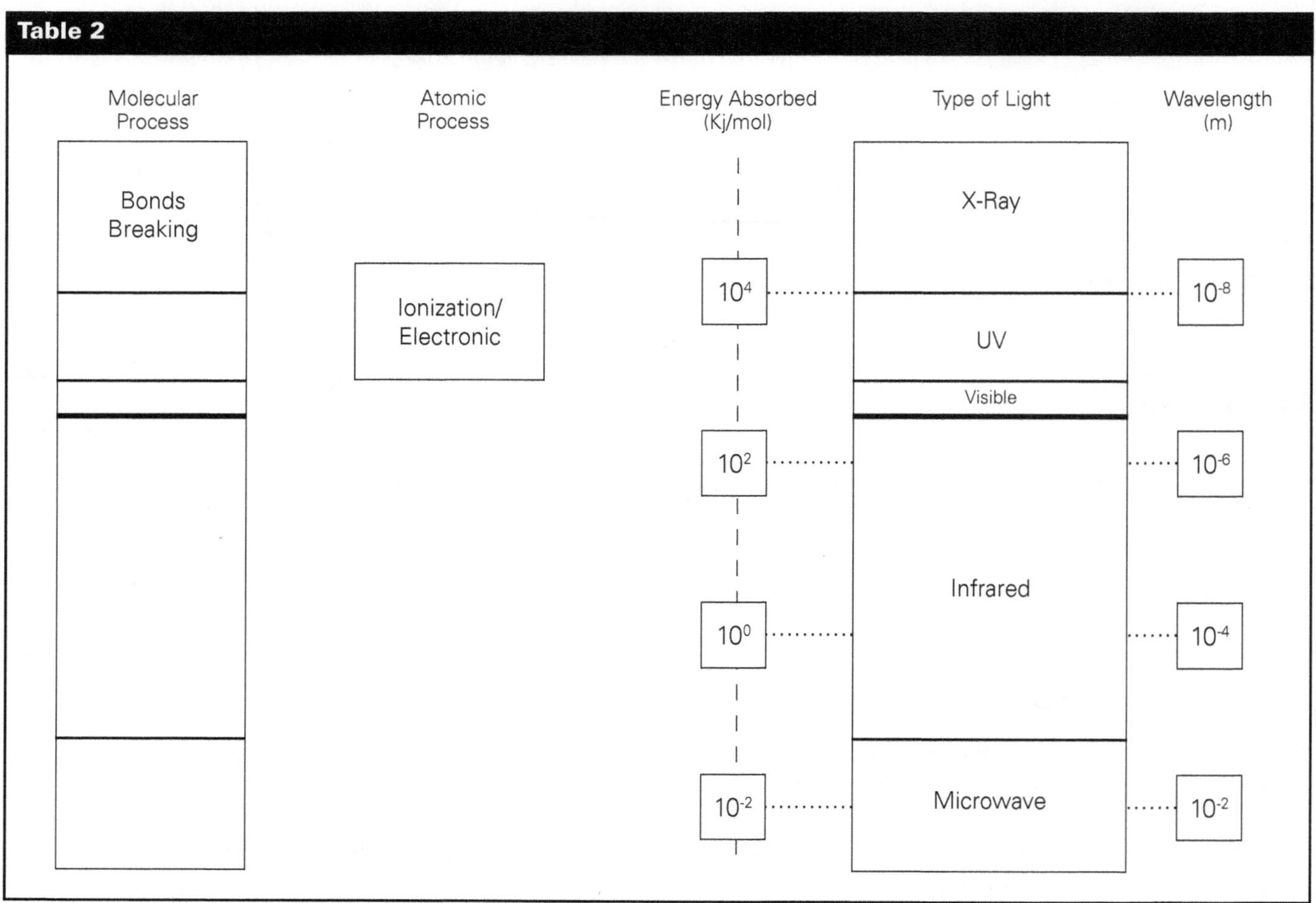

Consider this...

The different transitions in a molecule do not occur independently, but when enough energy is absorbed for an electronic transition, the vibrational and rotational transitions overlap with the electronic transitions. Since molecules can rotate and vibrate these transitions impact the spectrum. Let's assume that each electronic state of formaldehyde has 4 vibrational states. Each vibrational state would also have a number of rotational states, but the energy differences would be very small and hard to show on the diagram.

Figure 3 *Electronic Diagram*

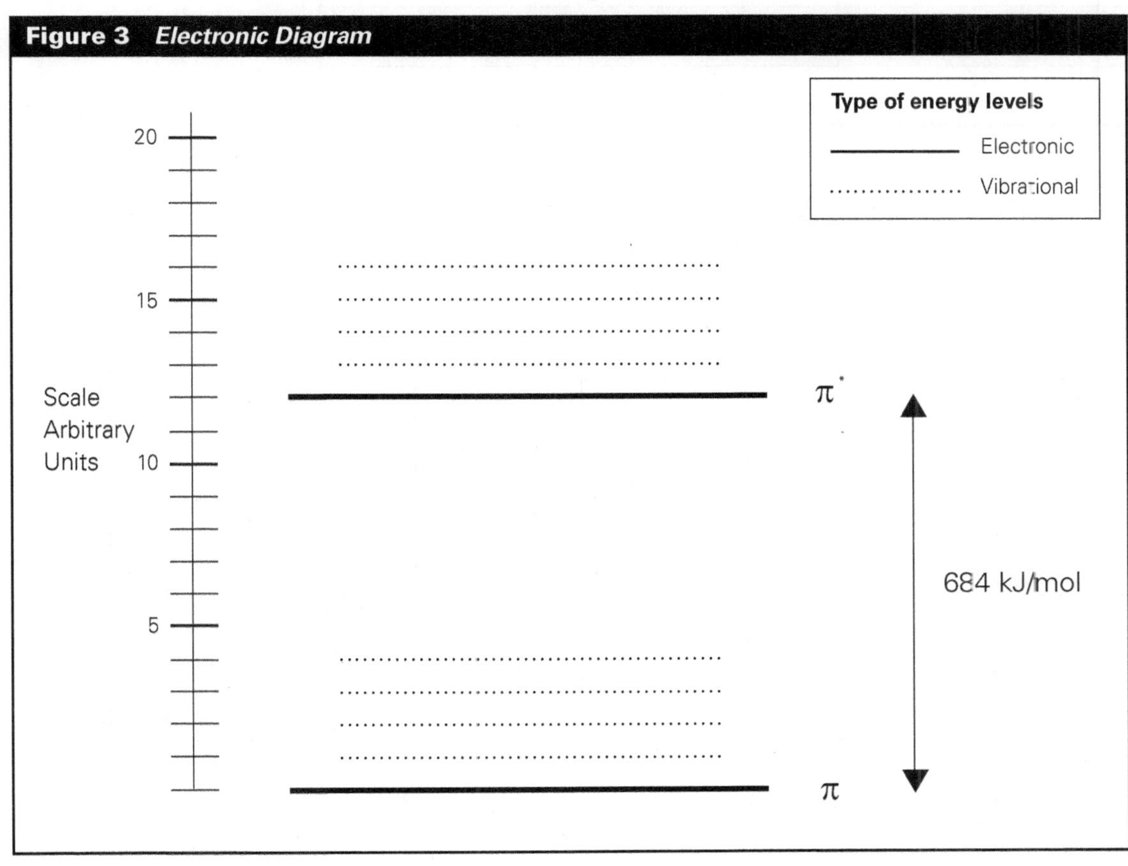

15. Draw all the transitions from one of the π energy levels to all the π* energy levels. Assign each student in your group to start from a different vibrational level in the π electronic energy level. (For this model, do not worry about "allowed" transitions)

16. Each transition has a length. Use the scale on the side to determine the length of each of your transitions and list those below.

________________ ________________ ________________

________________ ________________ ________________

17. For your entire group, determine the number of transitions of each length and record them in the table below.

Length	Number of Transitions	Length	Number of Transitions
8		13	
9		14	
10		15	
11		16	
12		17	

18. Sketch the length vs. number of transitions on the graph below and connect the points.

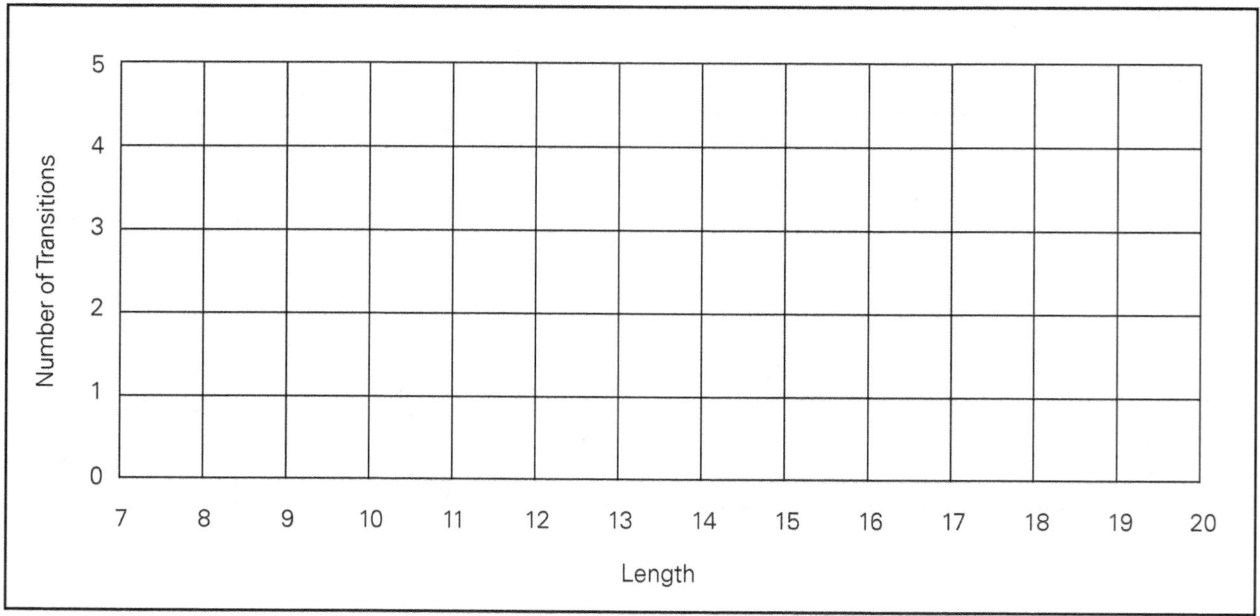

19. This sketch is analogous to one band in the UV-Vis spectrum for formaldehyde (see formaldehyde spectrum at the end of the activity if you need help). What does the length of each transition represent in the UV-Vis spectrum? What does the number of transitions represent?

The actual intensity at each wavelength is not due to only the number of possible transitions. Some transitions are more likely to occur and some have different intrinsic strengths.

20. Compare the spectra drawn in Q7 and Q18. Discuss with your group the differences and similarities between a molecular spectrum and an atomic spectrum. Record a list of both differences and similarities. Be sure to consider the range of possible wavelengths and the type of molecular processes recorded.

21. Examine the gas phase spectrum of formaldehyde and gas phase transmittance spectrum of sodium at the end of this activity and add any additional similarities and differences to your lists.

22. Based on molecular processes, discuss why a spectrum for molecular absorption is much broader than a spectrum for atomic absorption. Share your explanation with another group. Record a consensus explanation in your own words.

23. What are two specific concepts that you learned about molecular and atomic spectroscopy due to this activity?

24. Explain how this activity helped you to improve your ability to find similarities and differences between two items (i.e., spectra).

Application

25. Many descriptions of microwave ovens explain that the molecules vibrate and heat up. The typical microwave works at a frequency of 2.45 gigahertz (10^9 s^{-1}). Does the microwave radiation increase the vibrational motion of the molecules? (Note: $\nu\lambda = c = 2.998 \times 10^8$ m/s)

26. a. Calculate the wavelength of light needed for an atomic absorbance process requiring 405 kJ/mol of energy. ($h = 6.626 \times 10^{-34}$ J s, $c = 2.998 \times 10^8$ m/s)

b. In what region of the electromagnetic spectrum is this light found?

c. Most atomic absorbance processes between atomic orbitals occur in the UV/Vis region of the electromagnetic spectrum. Does the wavelength you calculated seem reasonable?

27. Alkanes such as n-hexane are used as solvents in UV spectroscopy. The energy difference for the $\sigma \rightarrow \sigma^*$ transition is 886 kJ/mol. What wavelength does hexane absorb? Why is n-hexane a good choice as a solvent for UV/Vis spectroscopy?

28. Is the following transmission spectra atomic or molecular? Explain your reasoning.

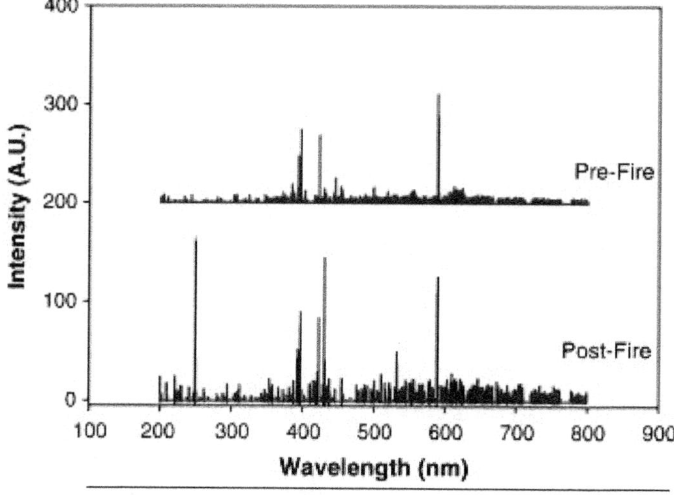

Used with permission: Spectrochimica Acta Part B: Atomic Spectroscopy, Volume 62, Issue 12, December 2007, Pages 1426-1432.

29. What are the differences in spectra between molecular and atomic absorption? Explain these differences using the type of transitions.

30. UV-Vis instruments used to measure molecular absorption typically have a broad wavelength light source while atomic absorption instruments use a light source that produces very narrow bands of light. Both are typically in the same UV-Vis wavelength range. Why don't they use the same source?

Appendix

UV-Vis Spectrum of Formaldehyde

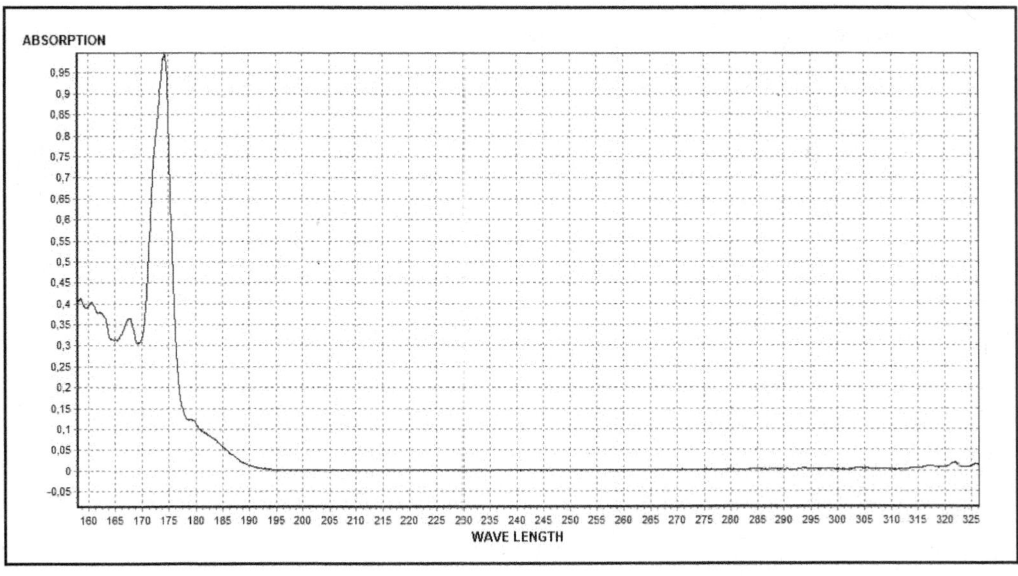

Used with permission: Chromalytica AB

Sodium Transmittance Spectrum

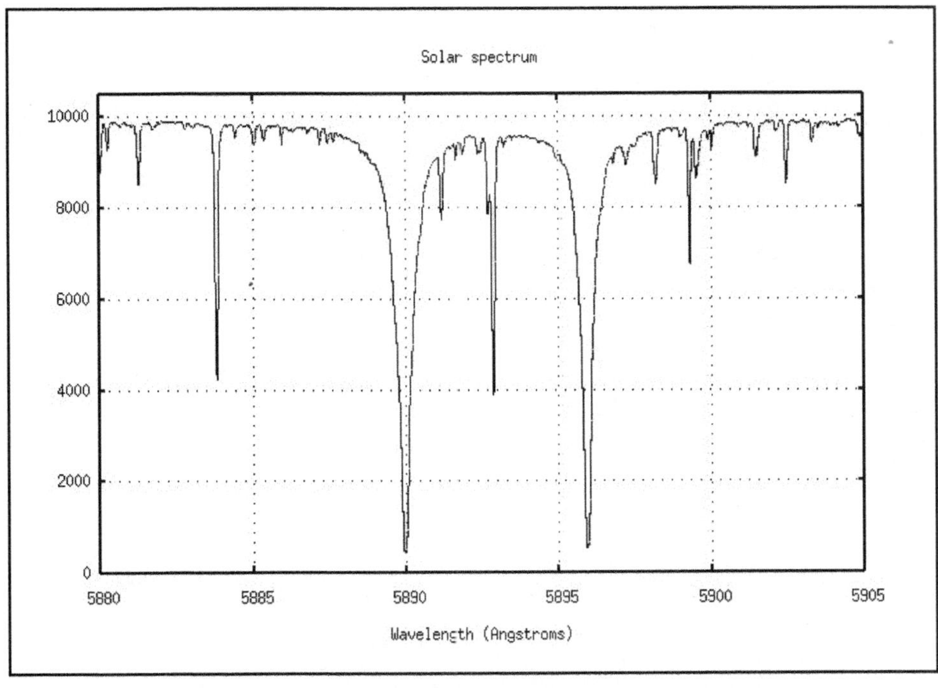

Used with permission: Michael Richmond, Rochester Institute of Technology

Infrared Spectroscopy: Vibrational Modes

Learning Objectives

Students should be able to:

Content

- Classify a vibration as a bend or a stretch.
- Predict the number and type of motions for a particular molecule.
- Explain the relationship between Infrared (IR) absorbance and changes in dipole moment.

Process

- Use the chemical education digital library to view molecular vibrations and corresponding IR spectra.
- Infer general relationships from specific examples (Generalizing).
- Specify the frequencies of some infrared vibrational modes (Integrating prior knowledge).

Prior knowledge

- Frequency, wavelength, wavenumber and their relationships.
- Ground state, excited state, and dipole moment.
- Absorbance, transmittance, and corresponding equations.
- Distribution of electrons in a compound using valence bond and molecular orbital theories.
- Interpretation of IR spectra of organic molecules.
- Familiarity with the electromagnetic spectrum.

Further Reading

- Harris, D.C., *Quantitative Chemical Analysis,* 8th Edition, 2010 W.H. Freeman: USA, Sections 17-1,17-6, and 19-5, pp.394-395, 404-407, and 467-472.
- Skoog, D.A., F.J. Holler, and S.R. Crouch, *Principles of Instrumental Analysis,* 6th Edition, 2007. Thomson: USA, Chapters 6 and 7, pp.430-480.
- Molecules 360, Chemical Education Digital Library, 2007. http://www.chemeddl.org/collections/molecules/index.php

- Reusch, William. "Virtual Textbook of Organic Chemistry," Infrared Spectroscopy Chapter, Michigan State University, 1999, http://www.cem.msu.edu/~reusch/VirtTxtJml/Spectrpy/InfraRed/infrared.htm has an excellent tutorial on topics related to this activity.
- Molecular Vibrations website by Edwin A. Schauble, Department of Earth and Space Sciences, UCLA, http://www2.ess.ucla.edu/~schauble/molecular_vibrations.htm
- Henderson, Giles; Liberatore, Christine. *Animated Vibrational Models of Triatomic Molecules,* J. Chem. Educ. 1998 75 779.
- Spectral Database for Organic Compounds SDBS, http://riodb01.ibase.aist.go.jp/sdbs/cgi-bin/cre_index.cgi?lang=eng

Authors

Christine Dalton, Mary Walczak, Carl Salter

Consider this...

The molecular models and infrared spectra for many small molecules can be viewed as animations at the Chemical Education Digital Library website: http://www.chemeddl.org/. Go to this website and choose the *Models 360* application. Type CO_2 into the Find: box and a ball-and-stick model of CO_2 will appear in the frame on the left.

On the right there is a tab called "Molecular Vibrations;" if you open this tab and select one of the entries in the list, the ball and stick model of CO_2 will move with the motion associated with that vibration. Clicking on "View IR Spectrum" will open a new window that shows a simulated IR spectrum of the molecule, displaying the predicted frequency and intensity of each vibration. You can also select a vibrational motion for the model by clicking on a peak in the IR spectrum. In both the vibration list and the spectrum, the frequencies of the vibrations are reported as wavenumbers, cm^{-1}.

Key Questions

1. Examine the vibrational spectrum of CO_2. Note the x-axis units of wavenumbers (cm^{-1}). Convert 640.236 cm^{-1} to nm and to kJ/mole. In what region of the electromagnetic spectrum does this wavelength fall?

2. Infrared spectra are typically displayed as graphs with x-axis values of wavenumbers, given the symbol. The units on wavenumbers are always inverse distance, and in the case of infrared spectroscopy are usually cm^{-1}. Given that energy is related as shown at the right, circle the relationship between wavenumbers and energy, frequency and wavelength.

$$E = h\nu = \frac{hc}{\lambda} = hc\tilde{\nu}$$

Wavenumbers are	*directly*	or	*inversely*	proportional to energy
Wavenumbers are	*directly*	or	*inversely*	proportional to frequency
Wavenumbers are	*directly*	or	*inversely*	proportional to wavelength

[1] Unfortunately, "View IR Spectrum" is not supported by Internet Explorer, so your browser may not be able to display the spectrum. If your browser does not support "View IR Spectrum," you can still complete this activity, but will not be able to determine the intensity of the vibration bands. Therefore, your "Transmittance" can not be determined and the sketches will have to approximate the spectrum. Firefox, Chrome and Safari browsers support the spectrum feature of the site.

3. Observe all the vibrations for CO_2. Sketch the IR spectrum of CO_2 in the space below on the left, taking special note of the values along the x-axis. If you are unable to view the spectrum, use the Normal Modes of Vibration (the numbers are wavenumbers (i.e., cm^{-1}) of that mode) to construct a spectrum. If you are unable to view the spectrum, focus on the position of the lines on the frequency axis. Compare and contrast your sketched spectrum with the experimental spectrum for CO_2 at the right.

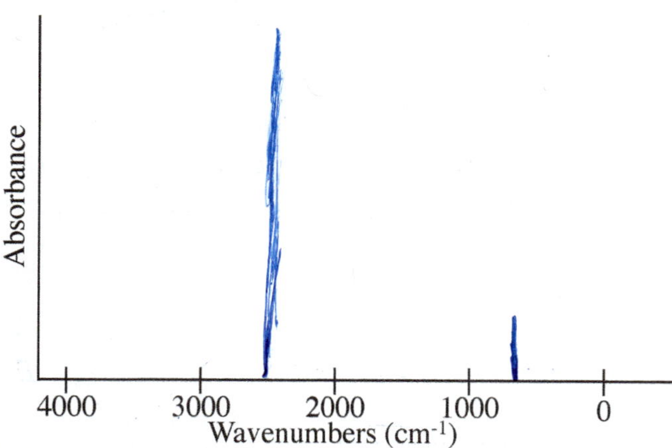

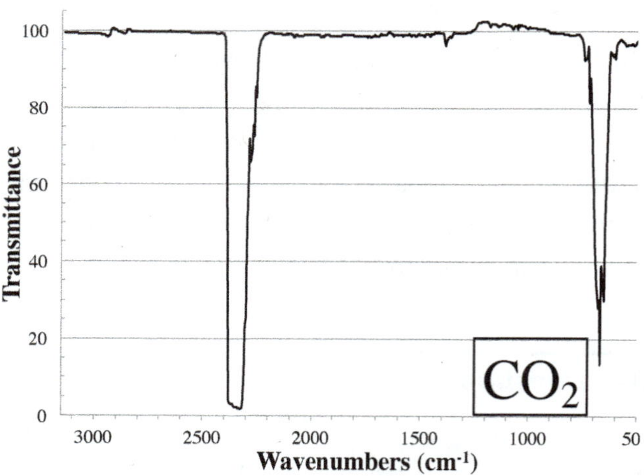

4. Using the animations of each vibration available on the website, classify the vibrational modes of CO_2 as stretches and bends. Make a list of the peaks and the corresponding frequencies in wavenumbers. Check for consensus among the group members.

Peak	Frequency (cm^{-1})	Type of Movement
1	640.236	bend
2	640.236	bend
3	1371.790	stretch
4	2435.848	stretch

5. How many vibrations are apparent in the experimental spectrum? How does the number of vibrations compare to the number predicted on the website? Based on the predicted spectrum and the experimental spectrum, which vibration is infrared inactive?

4 vibrations appear in the spectrum out of the predicted expected 4. Two overlap at 640.236 cm⁻¹, Based on both spectra, the vibration at

6. Sketch the IR spectrum of H_2O based on data from the Chemical Education Digital Library website. List the peaks and the corresponding energies in wavenumbers and label each as a bend or a stretch.

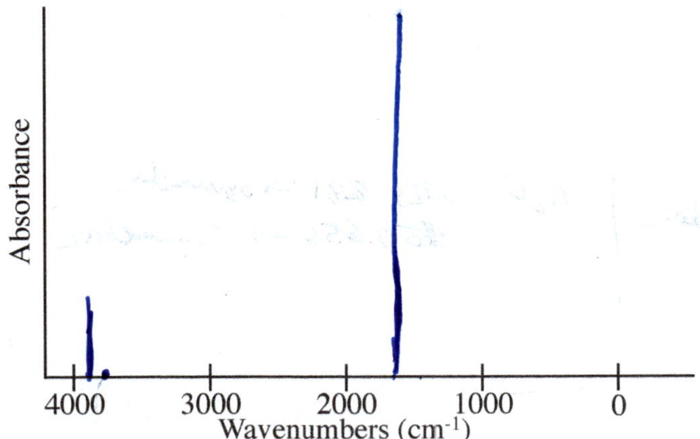

Peak	Frequency (cm^{-1})	Type of Movement
1	1713.782	bend
2	3729.281	stretch
3	3850.656	stretch

7. Sketch the IR spectrum of H_2O_2 based on data from the Chemical Education Digital Library website. List the peaks and the corresponding energies in wavenumbers and label each as a bend or a stretch.

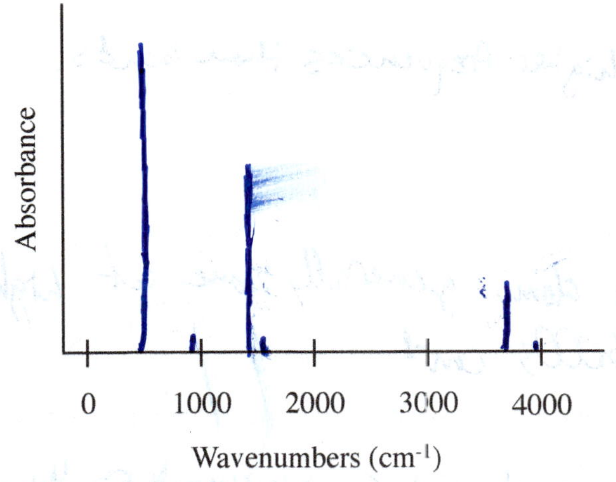

Peak	Frequency (cm^{-1})	Type of Movement
1	356.661	bend
2	957.511	stretch
3	1331.154	bend
4	1470.327	bend
6	3707.668	stretch
7	3710.356	stretch

Vibrational modes are classified according to movements such as symmetric stretching, asymmetric stretching and bending (sometimes called scissoring). A molecule's vibrations may be further classified as symmetric stretches (when the motion results in symmetric displacement of atoms) or asymmetric stretches (when the motion displaces atoms in an asymmetric way). Sometimes molecules have more than one vibrational motion that has the same energy. Such modes are called degenerate to indicate that the energy required to excite any of these vibrations is the same.

8. Look again at the peak tables constructed for CO_2 and H_2O in Q4 and Q6. Identify pairs of peaks that are:

A. Symmetric and asymmetric stretches.

CO_2: 1371.790 → symmetric
2435.848 → asymmetric

H_2O: 3729.281 → symmetric
3850.656 → asymmetric

B. Degenerate vibrations.

CO_2: 1 & 2 @ 640.236 cm^{-1}

Check your answers by comparing with another group.

9. Using the information collected in Q4, Q6, Q7 and Q8 for CO_2, H_2O and H_2O_2, make a general statement about:

A. The frequencies of bends vs. stretches.

Generally, stretches occur at higher frequencies than bends.

B. The frequencies of stretching vibrations involving hydrogen atoms.

stretches involving hydrogen atoms generally come at high frequencies, into the 3000s cm^{-1}

C. The intensity of stretching vibrations when two bonds appear to increase or decrease in length at the same time. (For example, the 1372 and 2436 cm^{-1} vibrational modes of CO_2)

asymmetric stretches come at higher wavenumbers than symmetric stretches.

Consider this...

The atoms in molecules can move in space in several different ways. For a molecule with n atoms, the total number of movements is 3n, corresponding to each of the atoms moving in three dimensions: x, y and z. All n atoms might move together in one direction resulting in a translation. The molecule can also rotate along the x, y and z axes. The remaining motions are vibrations resulting from the atoms moving independently. For example, the carbon dioxide molecule, shown in Figure 1 below, has a total of 3n=3(3)=9 distinct motions. Three of these are translational motions in the x, y and z directions. The rotations about two of the three axes (x and z in Figure 1(a)) result in movement of the atoms. The third rotation, in this case rotation about the y axis, does not change the position of the atoms, so it is excluded. The four final motions are the vibrations of the molecule, as shown in Figure 1(b). These can be better visualized on the Chemistry Education Digital Library by clicking the box "Show Vectors" underneath the list of Molecular Vibrations.

Figure 1 *Carbon dioxide. (A) Orientation in three-dimensional space. (B) Four possible vibrations.*

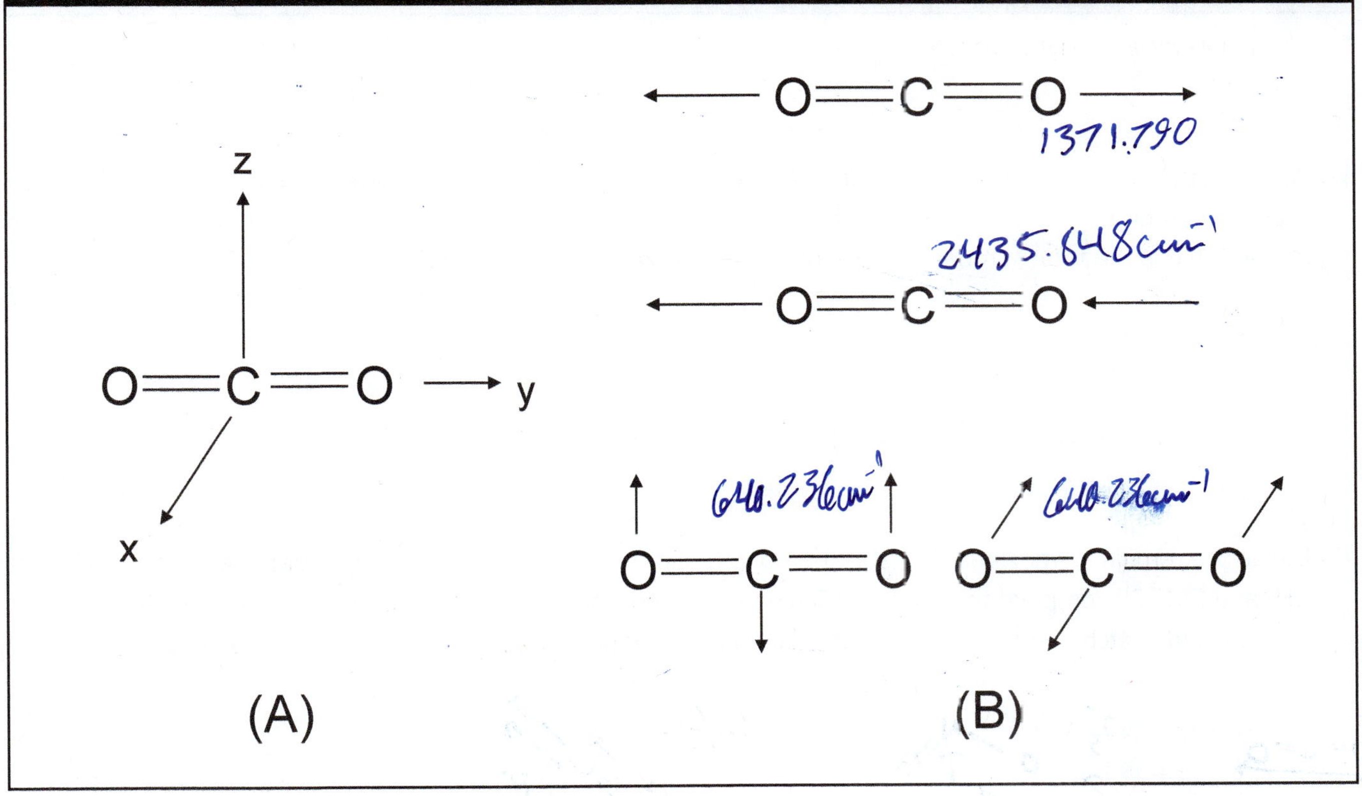

Key Questions

10. How many translations can the CO_2 molecule undergo? How many rotations? How many vibrations? Add your answers to Table 1 below. What is the total number of molecular motions that CO_2 can undergo?

Table 1 ***Summary Information on Molecular Motions***

	Carbon Dioxide	Water	Carbon Monoxide	Nitrate Ion	Methane	Dichloroacetylene
Number of atoms	3	3	2	4	5	4
Translations	3	3	3	3	3	3
Rotations*	2	3	2	3	3	2
Vibrations	4	3	1	6	9	7
Total	9	9	6	12	15	12

* result in movement of the atoms from original position

11. As a group consider the water molecule, which also contains 3 atoms. How many translations can the H_2O molecule undergo? How many rotations? How many vibrations? Add entries to Table 1 corresponding to water. What is the total number of molecular motions that H_2O can undergo

12. Divide the remaining molecules in Table 1 among group members. For each of the four molecules CO, NO_3^-, CH_4 and ClCCCl, complete the entries in Table 1. View the vibrational modes of the molecules on the website and sketch one symmetric stretch and one out of plane bend (as possible) for each molecule.

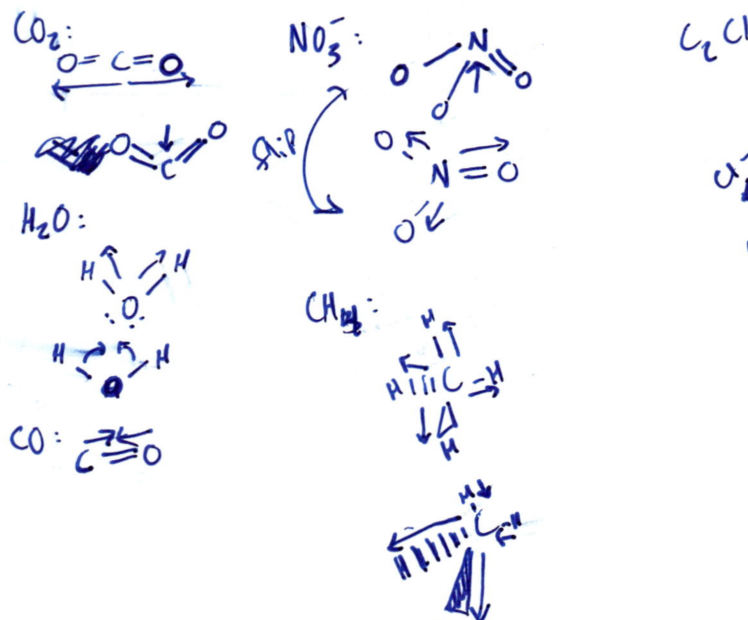

C_2Cl_2

13. Examine the entries in Table 1. What similarities and differences do you see for:

A. The total number of motions?

all multiples of 3

B. The number of translational motions?

all 3

C. The number of rotational motions?

no more than 3

D. The number of vibrational motions?

14. Write a mathematical expression for the number of vibrational modes of a molecule. Does the expression differ for linear and non-linear molecules? If so, explain.

Consider this...

Again, examine the IR spectrum of CO_2 below, or in the Chemical Education Digital Library.

Figure 2 *Infrared Spectrum of carbon dioxide, (left) and vibrational modes of carbon dioxide (right)*

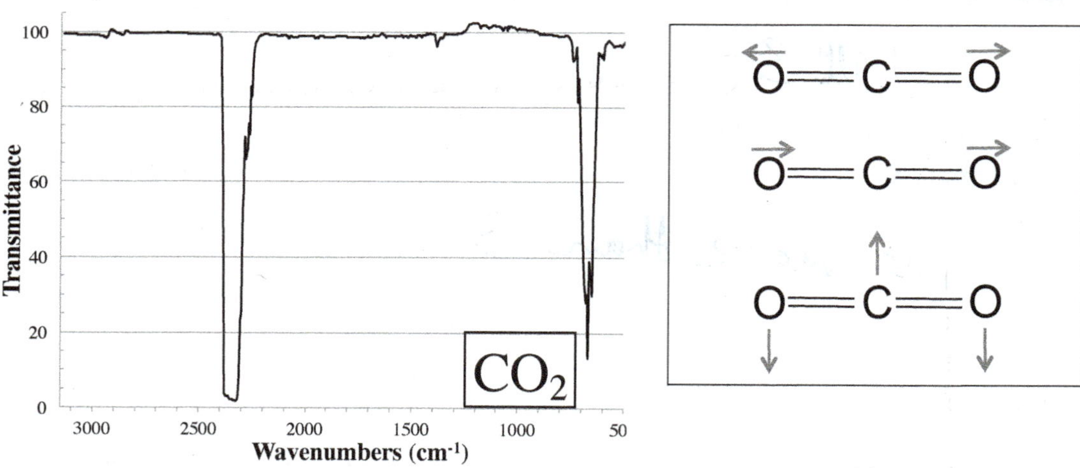

Key Questions

15. Look at the animated vibrations of CO_2 on the Chemical Education Digital Library website. For each of the four vibrational modes shown at the right (from Figure 1b), specify the frequency (in wavenumbers) corresponding to that vibrational motion.

Done pg. 199

16. The CO_2 molecule, although composed of polar C-O bonds, has a net dipole moment of zero, as illustrated at the right. Divide the four vibrational modes among group members. For each vibration, draw the molecule when the displacement due to the vibration is greatest and indicate the dipole moment of the molecule at that time. Compare answers among group members. For which vibrations does the dipole moment of the molecule change from its rest position, shown above?

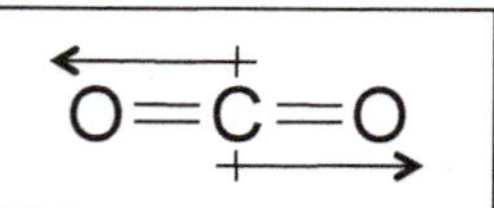

640.236 cm⁻¹:

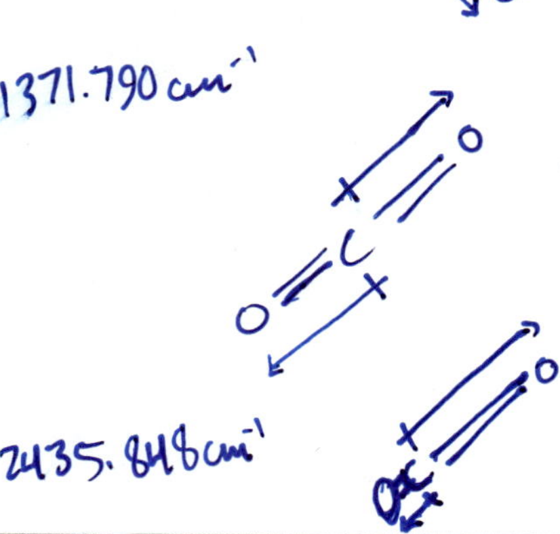

1371.790 cm⁻¹

2435.848 cm⁻¹

17. Using the information from your answers to Q15 and Q16, list reasons why the number of peaks in the CO_2 spectrum is inconsistent with the number of vibrations predicted in Table 1. Compare your group's answer to that of another group.

Only two peaks are visible because only two motions have dipole moments or changes in dipole moments

In order for a molecule to absorb infrared light two criteria must be met. First, the energy of IR light must be equal to the spacing between vibrational energy levels. Second, the vibrational motion must cause a change in dipole moment of the molecule.

18. Divide the following 4 molecules among group members: carbon monoxide, water, dinitrogen oxide, and nitrate ion. Does the molecule have a permanent dipole moment? Examine the vibrational modes for the molecule on the Chemical Education Digital Library website. Is there a change in the dipole moment when the molecule undergoes a vibrational motion? Will a vibrational mode result in an IR active absorption? Answer these questions in appropriate table below and share answers among group members.

Molecule	CO			
Permanent Dipole?	yes			
Vibrations (frequency)	2208.632			
Change in Dipole	yes			
IR Active?	yes			

Molecule	H_2O			
Permanent Dipole?				
Vibrations (frequency)				
Change in Dipole				
IR Active?				

Molecule	N_2O			
Permanent Dipole?				
Vibrations (frequency)				
Change in Dipole				
IR Active?				

Molecule	NO_3^-			
Permanent Dipole?				
Vibrations (frequency)				
Change in Dipole				
IR Active?				

19. Use the Chemical Education Digital Library website to find a molecule with one IR inactive mode and another molecule with two IR inactive modes.

1 IR inactive: N_2 (not at all IR active)

2 IR inactive: CS_2

20. Summarize what knowledge you have gained about vibrational modes of polyatomic molecules through this activity.

# of movements possible = 3 × # atoms. Translation in 3 directions is possible for all molecules, independent of identity.

21. List two connections you made between your understanding of infrared spectroscopy from organic chemistry and this activity.

1) Functional groups have specific ranges of wavenumbers because only certain ranges of energies are capable of making them vibrate.

2) Functional groups only appear in IR if they have a dipole moment or if a bend/stretch induces a dipole moment.

Application

22. Consider the molecule ammonia. Use the Chemical Education Digital Library website to examine the vibrational modes for the ammonia molecule. Sketch the vibrational modes and classify each as a bend or stretch. Specify the frequency of the infrared absorption for each of these modes.

23. Consider the oxygen molecule: O═O
Predict the number of vibrational modes for the oxygen molecule. Use the Chemical Education Digital Library website to examine the vibrational mode(s) for the oxygen molecule. Sketch the vibrational mode(s). Specify the frequency of the infrared absorption for each of these modes. Are the vibrational mode(s) for the oxygen molecule IR active?

24. Our atmosphere is composed of 78% N_2 and 21% O_2. Explain why it is not necessary to evacuate the sample compartment of an IR spectrometer.

25. Typically on a cloudy night the nighttime temperature does not drop as low as it does on a clear night. Explain why this is the case using the idea that clouds are made of water.

26. Carbon dioxide is classified as a "greenhouse gas." Explain what this means using the terms "IR active," "vibrational mode," and "radiate."

27. Consider the infrared spectra of acetic acid, ammonium ion, and ethane by viewing them on the Chemical Education Digital Library website. For each spectrum, indicate the vibrational modes associated with different spectral regions (e.g., O-H stretches).

28. Develop your own reference table of wavenumber ranges associated with vibrational motion using the Chemical Education Digital Library website.

(a) Functional Groups	**(b) Functional Groups**	**(c) Functional Groups**
O-H stretch	C-C stretch	C-O stretch
N-H stretch	C=C stretch	C=O stretch
C-H stretch	C=C stretch	C=O stretch

Introduction to Absorption Spectrophotometers

Learning Objectives

Students should be able to:

Content

- Identify the components of a UV-Vis spectrophotometer.
- Describe characteristics of the source, wavelength selector, sample holder, and detector in a UV-Vis spectrophotometer.
- Compare different types of instrument components based on these key characteristics and the goals of the instrumental analysis.

Process

- Interpret graphical data.
- Interpret schematic diagrams and relate to real instruments.
- Critical thinking by comparison and contrast.

Prior knowledge

- Atomic/molecular absorption and emission processes.
- Beer's law.
- Signals and noise.

Further Reading

- Harris, D.C. 2010. Quantitative Chemical Analysis, 8th Edition, pp. 445-460., New York: WH Freeman.
- Skoog, D.A., F.J. Holler, S.R. Crouch. 2007. Principles of Instrumental Analysis, 6th Edition, pp. 164-168, 175-203., Belmont, CA: Thomson Brooks/Cole.

Authors

Caryl Fish, David Langhus, and Ruth Riter

Consider this...

Figure 1 *Schematic of a UV-Vis spectrophotometer. Note that the dashed line between the detector and output represents the conversion of photons of light into an electrical signal and then to a number.*

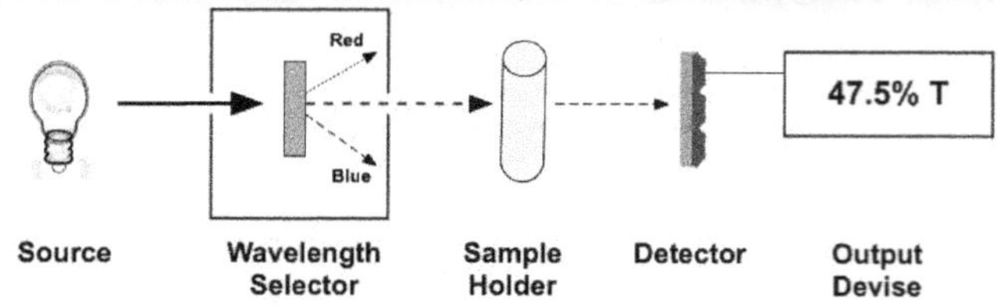

Key Questions

1. Examine Figure 1.

a. Following the arrows, how does the light intensity change as the light travels from the source to the detector?

b. Identify two aspects of the light entering the wavelength selector that differ from the light emerging.

c. Support or refute the following statement: The electromagnetic radiation absorbed by the sample reaches the detector.

d. Not all of the light from the source reaches the detector. What happens to this electromagnetic radiation? List at least three possible reasons for this loss.

Information: The detector converts the light transmitted through the sample into an electronic signal that is read by an output device most commonly a computer.

2. The electronic signal (typically current) has little meaning to the operator. What is the purpose of the output device?

Consider this...

Stability

Figure 2 ***Plots of intensity of emitted light at a given wavelength as a function of time for a candle and a fluorescent light bulb***

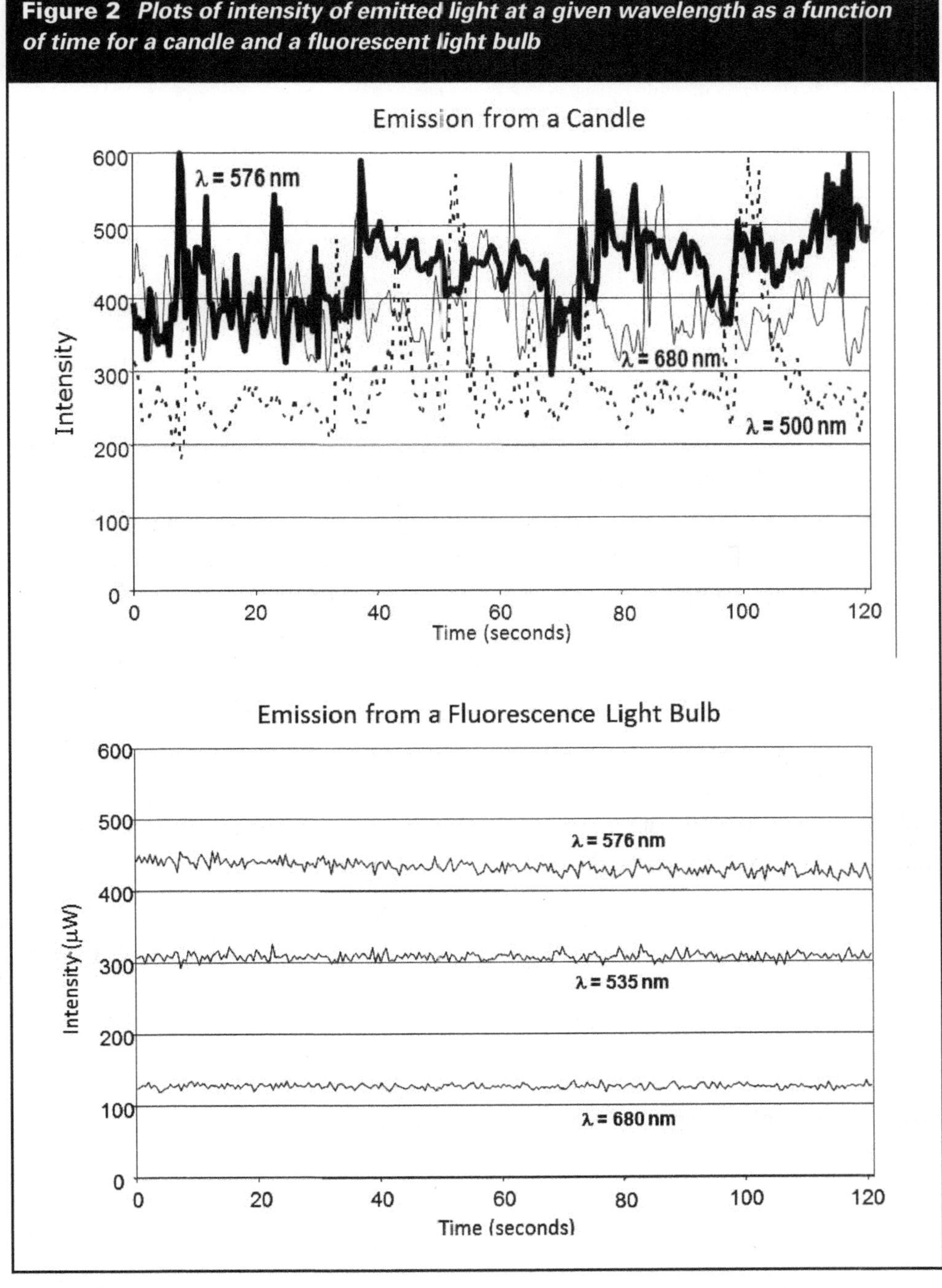

Key Questions

3. Examine the emission signals for the fluorescent bulb in Figure 2. At which wavelength is the emission signal more stable over time? (In other words, at which wavelength is the source's signal less noisy?) What evidence supports your answer?

4. Now compare the emission signals from both sources in Figure 2. Which one of these sources has more noise (or is less stable)?

Spectroscopists often quantify the stability of a source using a figure-of-merit called signal-to-noise ratio (S/N). The S/N is calculated by dividing the signal's mean by its standard deviation. When signals are given in graph form as in Figure 2, the mean is estimated by eye, and the standard deviation is estimated at the 99% confidence level by dividing the difference between the maximum and minimum signals by five.

SN=mean/standard deviation; Standard deviation~(Max Signal–Min Signal)/5

5. For the 576 nm emission signals of the candle and bulb, divide work among group members and estimate the average signal and standard deviation, then calculate the S/N using your estimations.

6. Write a concise statement describing the relationship between signal stability and S/N.

7. Let's now assume that the light source in Figure 1 could be either source shown in Figure 2. Recall that UV/Vis spectrophotometers measure P_0 and P and output $\%T$. Transmittance, T, is defined as the fraction of light transmitted by the sample: $T = P/P_0$ where P_0 is the power of the light before the sample holder and P is the power of the light after the sample.

   a. *Support or refute:* The source stability has an impact on P_0 and not on P.

   b. In practice, P_0 is measured then P is measured some time later. Discuss with your group members how the noise in the source affects the measured value of $\%T$.

8. Describe the relationship between signal stability and the quality of the absorption spectrophotometer's output signal.

Intensity

Key Questions

9. Using the emission data from Figure 2, answer the following:

   a. Which wavelength is the light most intense for each source?

   b. Remember the relationship between wavelength and energy. For the four wavelengths in Figure 2, which wavelength is the most energetic?

   c. *Support or refute:* The energy of a photon of light depends on the intensity of the light.

10. How might a low intensity source be problematic in a spectrophotometer?

Consider this...

Range

Figure 3 *Emission spectra of incandescent (left) and fluorescent (right) light bulbs*

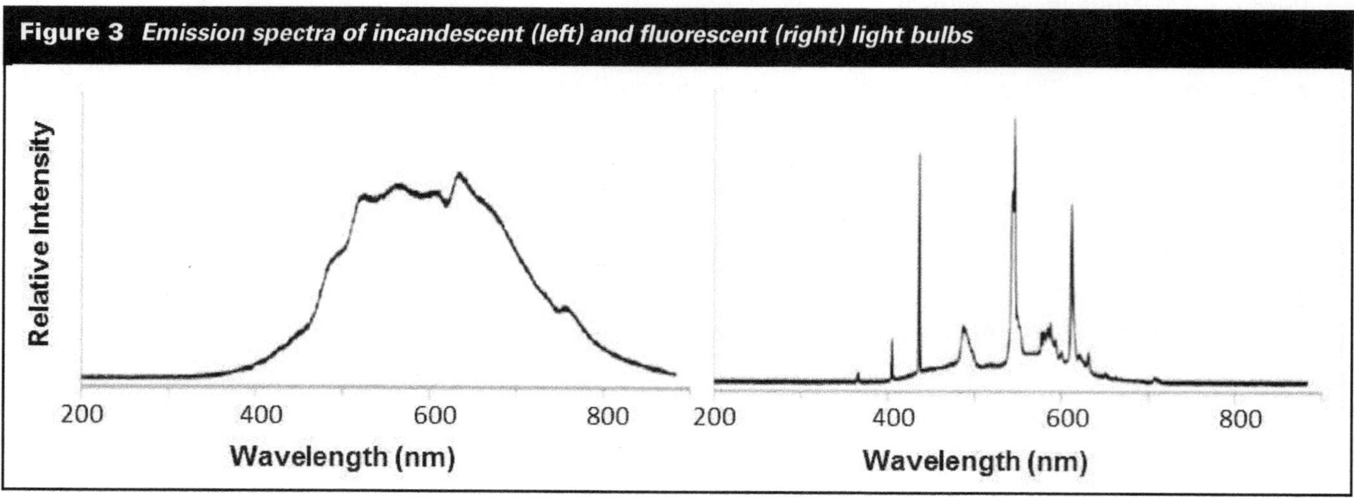

Key Questions

11. Over what wavelength range would each light bulb be a source of light for an absorption measurement?

12. Describe how intensity changes as wavelength increases for each source. In other words, are the intensities uniform?

13. At 525 nm, which is the better light source? State your evidence.

Continuous light sources change smoothly from one wavelength to the next. On the other hand, line sources produce narrow bands or lines of intense radiation.

14. Describe how an incandescent light bulb can be a continuous source in the visible region of the electromagnetic spectrum.

15. Would a fluorescent light bulb be a better continuous or line source? Explain your reasoning?

16. If you were buying a source, what information would you want a manufacturer to tell you so you can compare sources? List at least four criteria.

Consider this...

Ideally, when making an absorption measurement, the light passing through the sample is of a single wavelength or monochromatic. Polychromatic light causes deviations from Beer's law when calibrating the instrument. While it is not possible to isolate a single wavelength from a light source, it is highly desirable that the radiation is over as small a range of wavelengths as possible. In designing a general purpose spectrophotometer, it is also desirable that the source or sources provide(s) the largest range of wavelengths. A monochromator is a device reconciles these two conflicting attributes. Another important attribute of a monochromator is its throughput or light-gathering power. A large light-gathering power is desirable to insure sufficient sensitivity at the detector.

Figure 4 ***Schematic of a Czerney-Turner grating monochromator***

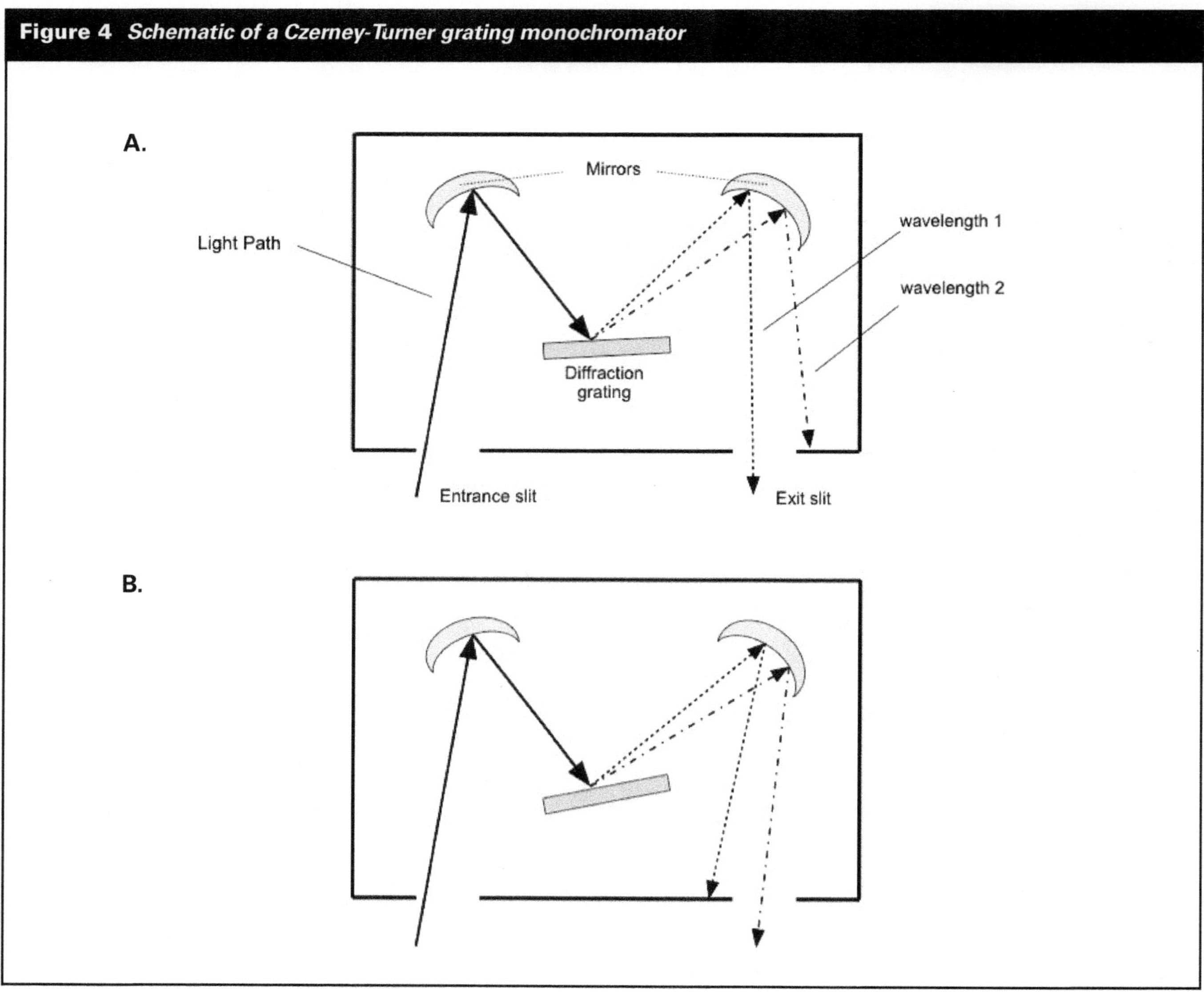

Key Questions

17. Based on the schematic shown in Figure 4, what is the function of the following components in the monochromator?

a. Diffraction grating

b. Mirrors

c. Exit slit

18. Compare the diagrams in parts A and B in Figure 4. How are they different?

19. How do the components change in order to select a wavelength to exit the monochromator?

Originally monochromators utilized prisms to separate the source's wavelengths. Modern monochromators use diffraction gratings as the dispersive element. Gratings separate the wavelengths of light through constructive interference, and as a result, different wavelengths focus at different places along the focal plane. This process is called *diffraction.*

Consider this...

Figure 5 ***Schematics of the Diffraction Grating, Mirror, and Exit Slit in a Monochromator. In parts B-D changes have been made to these components. Each line type (solid, dashed, and dotted) indicates a different wavelength of light. Keep in mind that light is continuous. These are just indicating where the wavelengths are located.***

A.

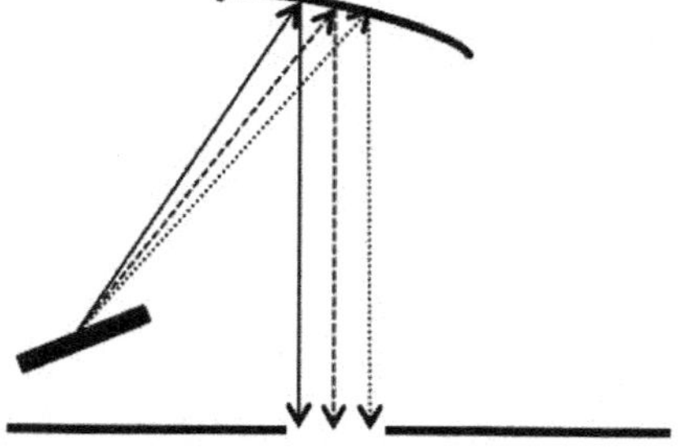

B.

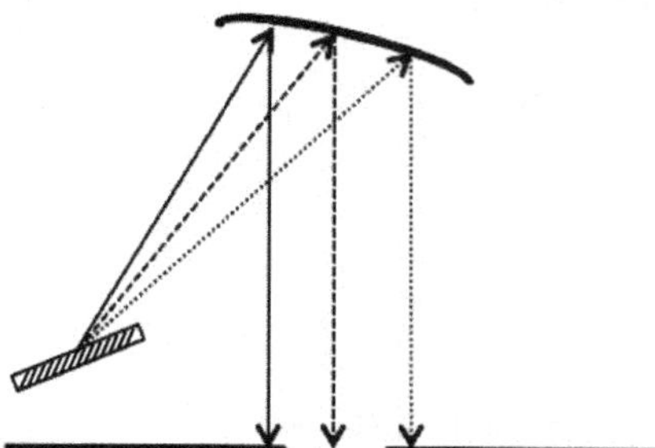

C.

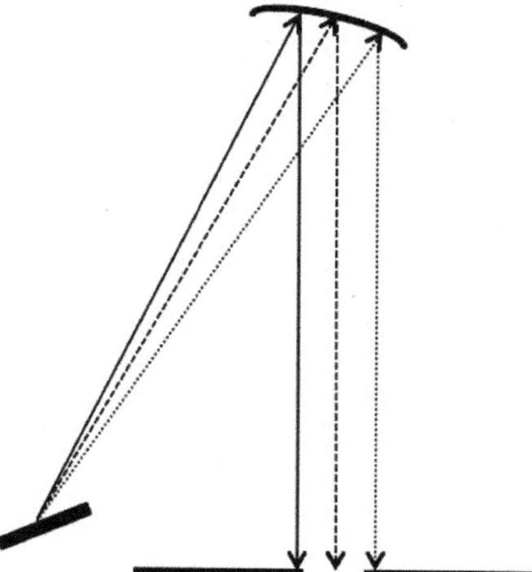

D.

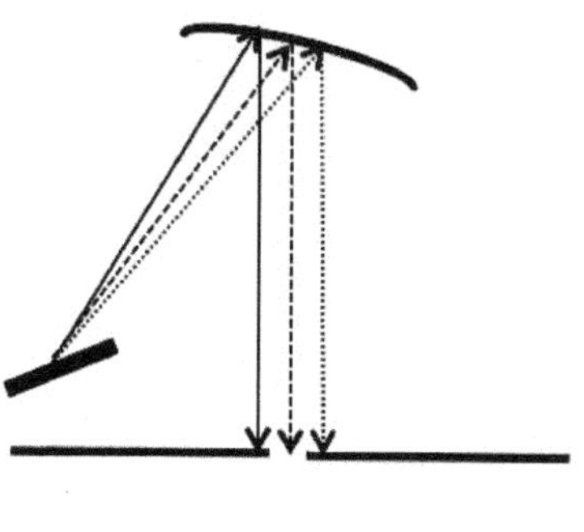

Key Questions

20. Examine Figure 5. Determine how the monochromator components differ in parts B-D from part A.

B.

C.

D.

21. Describe how these changes impact the resolution or the ability to separate one wavelength from another.

22. Narrowing the exit slit width would seem like the easiest way to improve resolution. However, narrowing the slit has a negative impact. Discuss with your group what this impact might be.

23. Summarize the purpose and design of a monochromator. Include the terms resolution and throughput in your description.

Part III: Sample Holders

Consider this…

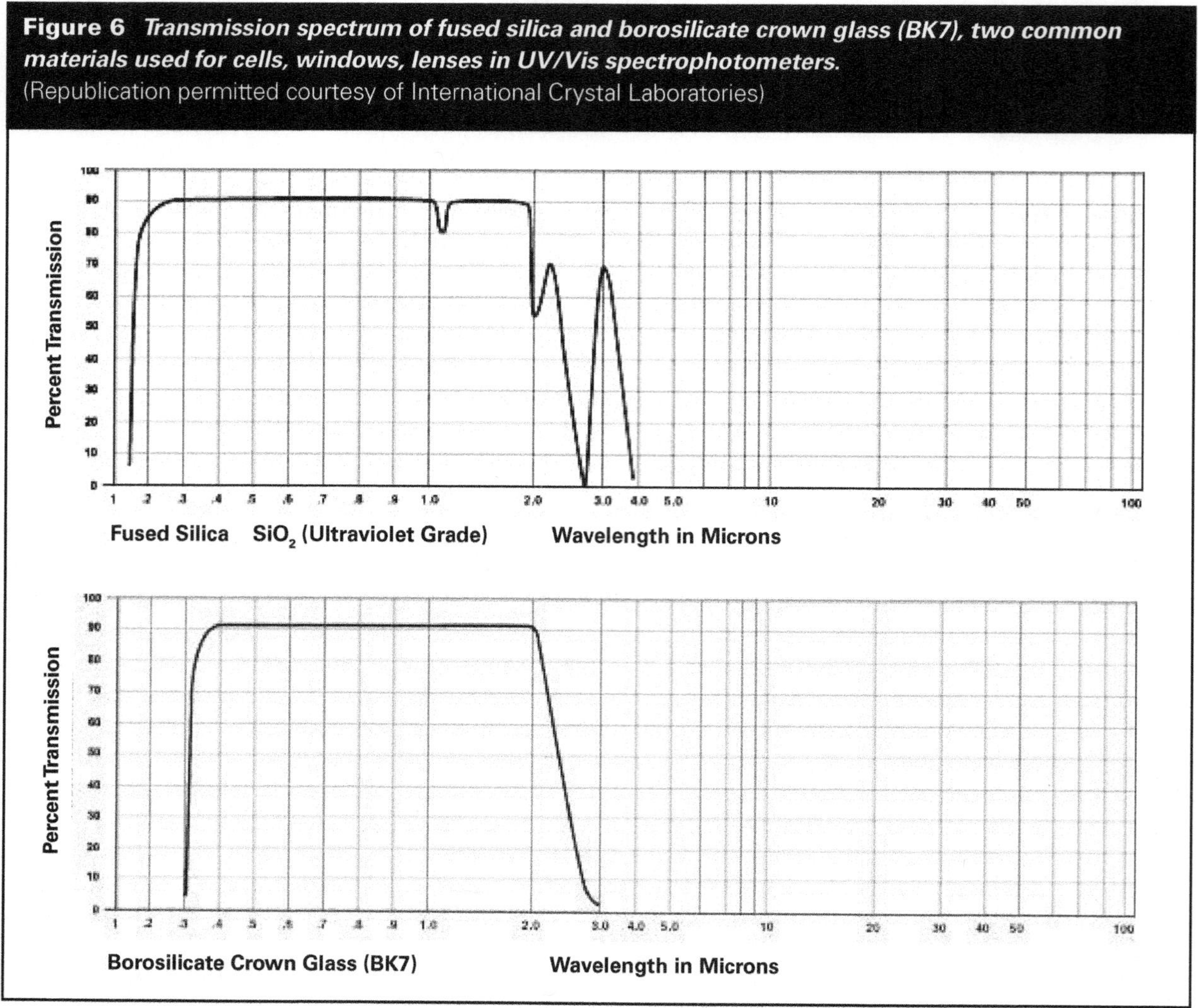

Figure 6 ***Transmission spectrum of fused silica and borosilicate crown glass (BK7), two common materials used for cells, windows, lenses in UV/Vis spectrophotometers.*** (Republication permitted courtesy of International Crystal Laboratories)

24. The UV/Vis region of the electromagnetic spectrum is from 200-800 nm. Label this region on each plot in Figure 6. (NOTE: microns are another word for micrometers, μm)

25. Examine Figure 6. What wavelengths of light are not absorbed by fused silica (i.e. they are transmitted), and why would fused silica be a good material to use in a sample holder for a UV/Vis spectroscopy experiment?

26. What wavelengths of light transmit through borosilicate crown glass (BK7), and at what wavelengths would this material not be useful for a UV/Vis absorption measurement?

Part IV: Detectors

Consider this...

There are several types of detectors used in UV/Vis spectrometers including photodiodes and photomultiplier tubes. As previously mentioned at the beginning of the activity, the detector converts ultraviolet or visible light into an electronic signal. A photodiode is a semiconductor device that allows current to pass when exposed to light, ideally one electron per photon. A phototube generates the electronic signal when light hits a photosensitive, negatively-charged surface, the cathode, and the electrons flow through a vacuum to a positively-charged collector, the anode. The resulting current is proportional to the number of photons hitting the surface of the detector.

Key Questions

27. From the above description of a phototube, label the following schematic of a photodiode.

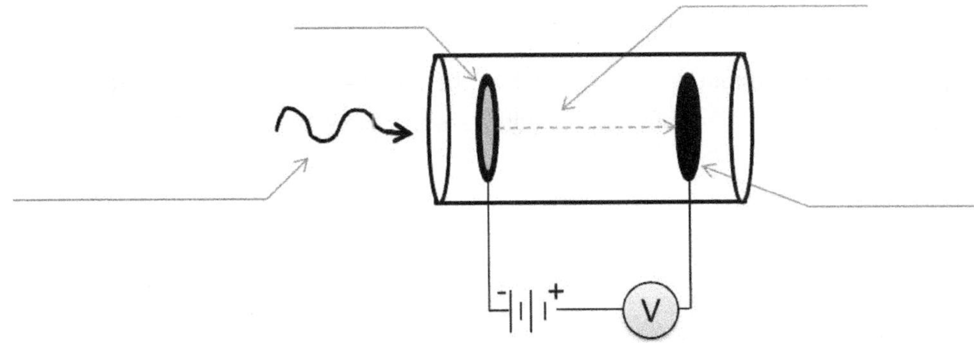

Consider this...

Photomultiplier tubes are phototubes that contain additional electrodes called dynodes that have a more positive voltage than the cathode resulting in the electrons accelerating as they move through the tube. In addition each electron that hits a subsequent dynode causes several additional electrons to be emitted.

Figure 7 ***Schematic of the cross section a photomultiplier tube.***

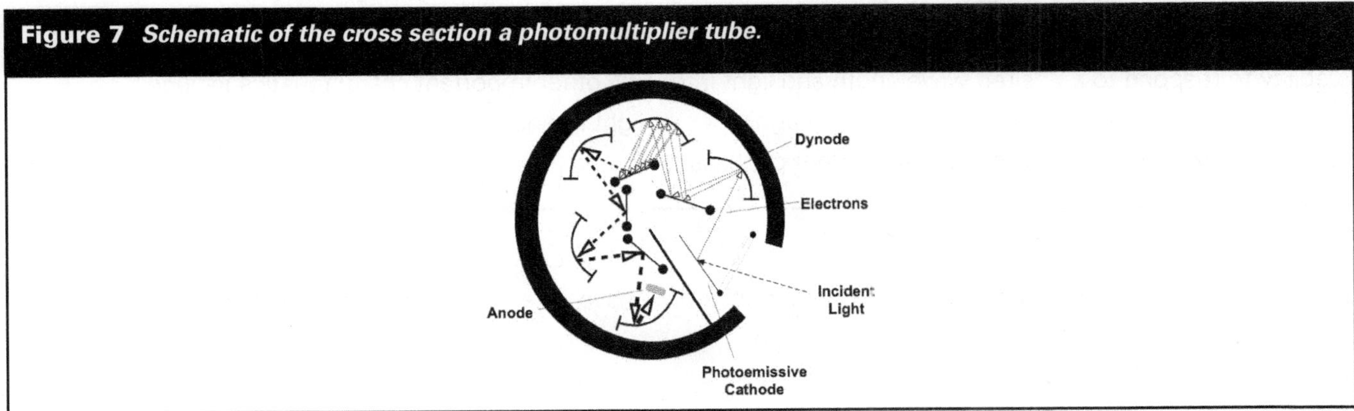

Key Questions

28. In Figure 7, consider the schematic on the right. When light (labeled incident radiation) hits the cathode, the cathode releases an electron. If there are 9 dynodes and we assume each electron hitting a dynode causes 2 electrons to be released, how many electrons would be released from the 9th dynode to the anode?

29. Electrons "flow" to the anode. What is the driving force for this movement?

30. The pressure inside the tube is very low (or in other words is a vacuum.) What is the advantage of this design?

31. Photomultipliers are several orders of magnitude more sensitive than a single phototube. Discuss what sensitivity means in this case, then determine why photomultipliers are more sensitive than a phototube and explain your reasoning.

32. Because of photomultipliers' sensitivity, the photomultipliers are particularly susceptible to "dark noise." Dark noise may be due to standard instrumental noise or stray radiation. Stray radiation is any light that hits the detector that did not go through the sample. Discuss with your group possible sources of stray radiation in a spectrophotometer and suggest possible solutions to reducing these sources.

> There are many desired requirements of a detector for UV/Vis spectrophotometers. In addition to a detector's ability to respond to a desired wavelength and light intensity, other important characteristics include a fast response time, linear response over magnitudes of light intensities, and high quantum efficiency (i.e., the percentage of electrons per incident photons.)

33. What effect would a detector with a slow response time have on an absorbance measurement?

34. Brainstorm in your group and determine what it means to say the detector has a linear response.

35. Why would a linear response of the detector over magnitudes of light intensities be an instrumental advantage?

36. Why would high quantum efficiency be a desired characteristic of a detector?

37. If you were buying a detector, what are four specific characteristics of the detector you would want a manufacturer to tell you?

38. Describe two concepts about spectrophotometers that your group is now more comfortable with.

39. Describe how your group is better able to interpret schematic diagrams of instruments.

Application

40. Describe each of the following spectra as either line or continuous and explain your reasoning. (Used with permission of B. Niece.)

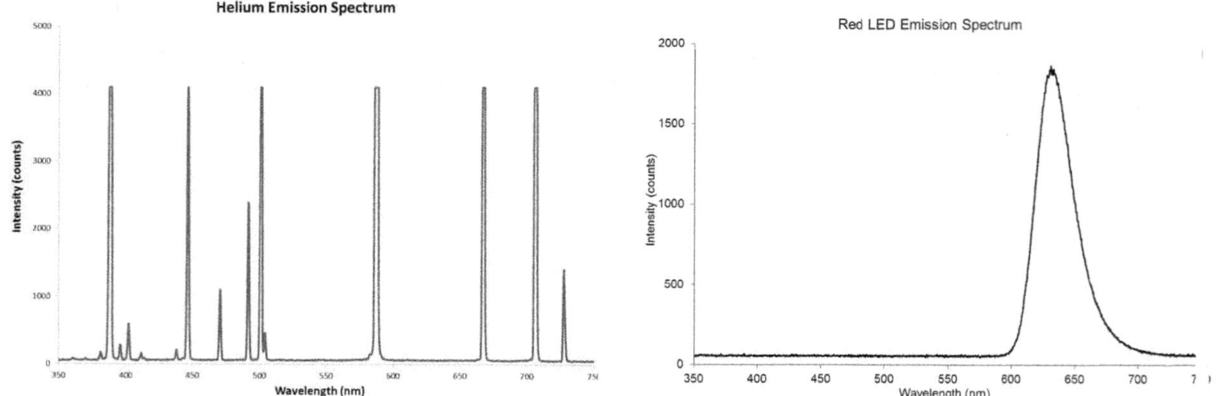

41. Laser has great stability and intensity but what about range? What are the advantages and disadvantages of using lasers?

42. A common way to determine the amount of chlorine residual in drinking water or swimming pools or spas involves observing the product of the reaction of aqueous chlorine with N,N-diethyl phenylenediamine reagent. Here's the absorption spectrum of that product:

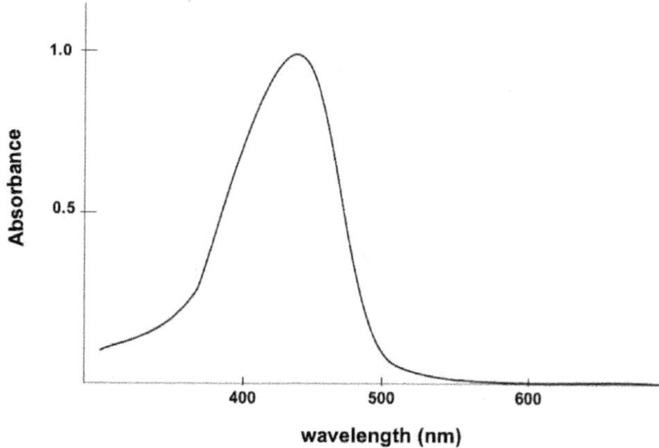

Determine what color(s) of light this sample absorbs. What is the color of the solution?

43. Notice from the UV-Vis spectrophotometer diagram in Figure 1 that the amount of light at the detector is significantly less than the source output. Would a more intense source be advantageous? Hint: think about signal-to-noise and explain your answer.

44. Some instruments actually employ more than one source. If you wanted to design an instrument that was capable of operation from 200 nm to 900 nm what sources would you choose? Justify your selection.

45. Disposable spectrophotometer cells eliminate cross contamination and the need for cleaning. Polystyrene and polymethyl methacrylate ("PMMA") cuvettes are available. At what wavelengths are these materials "transparent"? Why would you choose polystyrene over PMMA?

46. With your instructor's permission, examine a spectrophotometer in your department.

a. What is the light source for the instrument?

b. What kind of monochromator does the spectrophotometer have and how do you know?

c. How do you think wavelength is adjusted inside this instrument?

d. Is this a single beam or double beam instrument?

e. What is the detector in the instrument?

Mass Spectra Interpretation

Learning Objectives

Students should be able to:

Content

- Explain the source of fragment ions in electron impact mass spectra.
- Identify the base peak and parent (molecular ion) in a mass spectrum.
- Recognize patterns in the mass spectra and use them to identify functional groups.

Process

- Comparing and contrasting graphical data (critical thinking)
- Analyze graphical information (Information processing)

Prior knowledge

- Basic knowledge of mass spectroscopy instruments

Further Reading

- Organic Textbook's spectroscopy chapter.
- University of Arizona Tutorial (http://www.chem.arizona.edu/massspec/)
- NIST Mass Spec Data Center, S.E. Stein, director, "Mass Spectra" by in NIST Chemistry WebBook, NIST Standard Reference Database Number 69, Eds. P.J. Linstrom and W.G. Mallard, National Institute of Standards and Technology, Gaithersburg MD, 20899, http://webbook.nist.gov, (retrieved May 23, 2013).

Author

Caryl Fish

Consider this...

Mass Spectrometry examines the ions produced from energized molecules. Often this occurs after bombarding the molecules with high energy electrons. This electron impact dislodges electrons from the molecules, which become positive ions. Additional energy from the impact breaks one or more of the chemical bonds and forms fragments of the molecules. When a molecule fragments, it produces a charged fragment and a neutral molecule. Only the charged fragments are shown in the table. The cations are separated in the mass spectrometer based on their mass-to-charge ratio *(m/z)*, and a mass spectrum is produced showing the relative abundance vs *m/z*. The spectrum is the result of the fragmentation of a huge number of molecules.

Table 1 ***Possible ions formed from 2-butanone and their m/z.***

Bond Breaking	Fragment 1	*m/z*	Fragment 2	*m/z*
$CH_3CH_2C(=O)CH_3^+$	$CH_3CH_2C(=O)CH_3^+$	(72)		0
$CH_3CH_2C(=O)\vert CH_3^+$	$CH_3CH_2C(=O)^+$	(57)	CH_3^+	(15)
$CH_3CH_2\vert C(=O)CH_3^+$	$CH_3CH_2^+$	(29)	$CH_3\text{-}C(=O)^+$	(43)
$CH_3\vert CH_2C(=O)CH_3^+$	CH_3^+	(15)	$CH_3C(=O)CH_2^+$	(57)
$CH_3CH_2C(=O)CH_3^+$	$H_3CCH_2\overset{+}{C}CH_3$	(56)	O^+	(16)

Key Questions

1. What is the mass of the unfragmented (parent) ion? The mass of the $CH_3CH_2^+$ fragment ion?

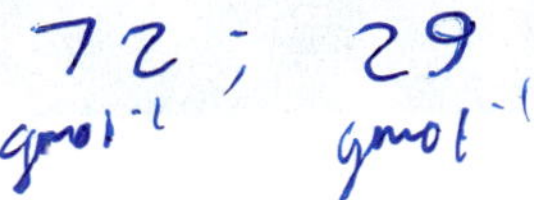

2. What is the charge on all these ions?

cationic, +1

3. For the ions in Table 1, how does the *m/z* ratio compare to the mass of the fragments?

they're the same

4. From the description of mass spectroscopy, after bombardment by the high energy electron, how do the parent ion and fragment ions form from the molecule?

a high-energy projectile ruptures the bond.

5. Complete Table 1 by adding the final ions with the masses of 56 and 16.

6. Plot below the number of fragments formed for each *m/z* in a bar graph.

Model of Mass Spectra

Model of Mass Spectra

Abundance: 0, 1, 2

m/z: 0, 10, 20, 30, 40, 50, 60, 70, 80

Consider this...

Figure 1 ***Actual Mass Spectrum of 2-butanone***
(Used with permission of National Institute of Standards and Technology)

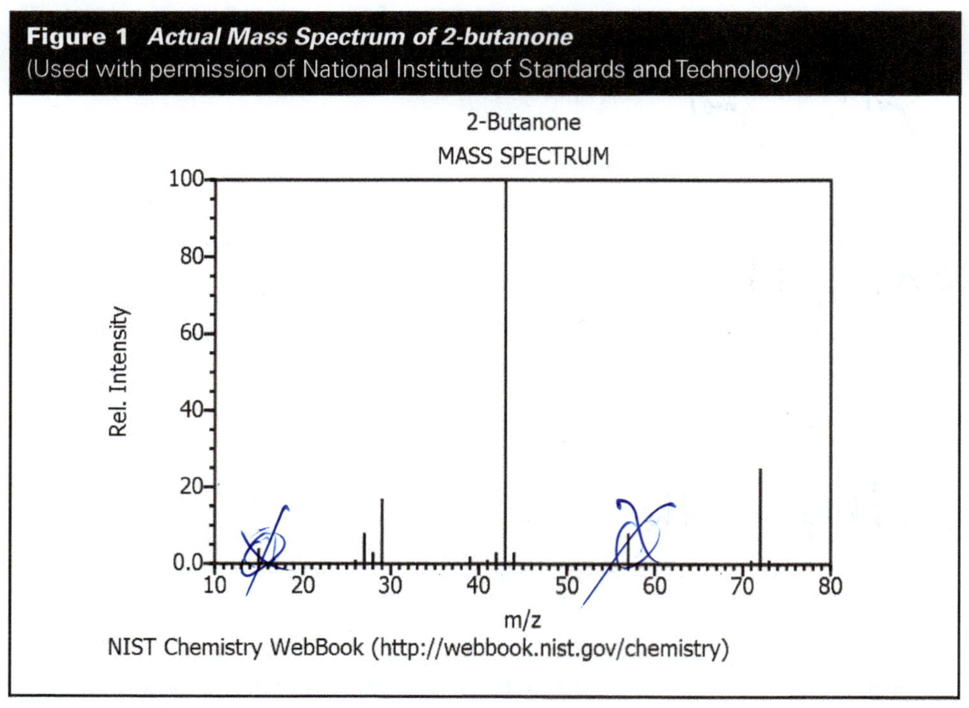

Key Questions

7. What are the *m/z* of the three largest peaks in the mass spectrum of 2-butanone?

29, 43, 72

8. Individually find two similarities and two differences between your model mass spectrum and the actual spectrum.

Sims
1) ~~[illegible]~~
2) same scale

Diff's
1) More peaks in Figure 1
2) Differences in peak height

9. Compare your list with your group. Develop a comprehensive list of similarities and differences between the two spectra.

10. Which of the possible ions in Table 1 are not found in the actual mass spectrum?

16 and 56 m/z

11. Specifically, why do think the 56 *m/z* peak is not in the mass spectrum?

The C=O is way less likely to break than any of the C-C bonds.

12. The abundance of a peak in a mass spectrum is often due to the stability of the ion. Based on the mass spectrum from 2-butanone, which ion is the most stable?

43 m/z

13. Notice the tiny peaks at *m/z* 73 and 44. Peaks one (m+1) or two (m+2) mass units larger than a major peak are often due to isotopes. What isotopes of carbon and/or hydrogen would account for the m+1 peaks?

^{13}C and ^{1}H

14. The ion with a mass of 72 in the 2-butanone spectrum is known as the molecular ion or parent ion. Why is it given this name?

It is the ion from which all other fragments derive

15. The base peak has the largest abundance. In some mass spectrometers the other peaks are compared to it to get relative abundance. What is the base peak in the 2-butanone spectrum?

43 m/z

16. As a group, list three reasons that the actual mass spectrum is different from the predicted spectrum.

1) Didn't consider ion stability
2) Didn't consider isotopes
3) 16 and 56 m/z don't actually appear; C=O doesn't break

Consider this...

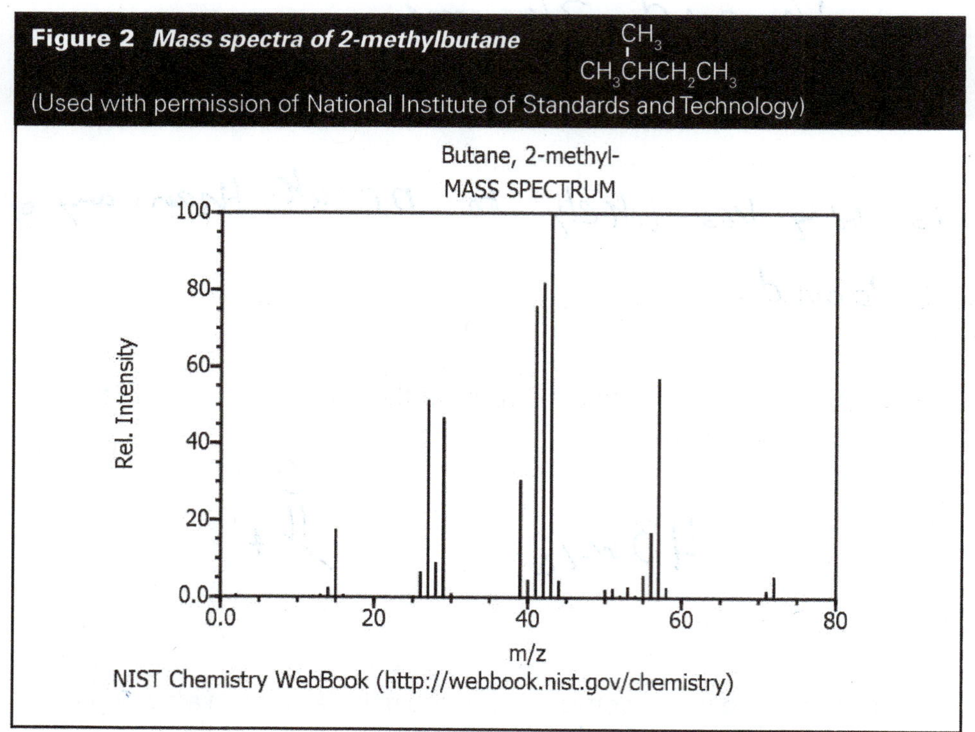

Figure 2 *Mass spectra of 2-methylbutane* $CH_3CH(CH_3)CH_2CH_3$

(Used with permission of National Institute of Standards and Technology)

72 g mol^{-1}

Key Questions

17. What are the parent (molecular) ion and base peak for 2-methyl butane?

72 m/z

18. Sometimes it is helpful to examine the groups that are leaving by subtracting the *m/z* of the fragment from the *m/z* of the molecular ion. What is the difference in *m/z* between the molecular ion and the base peak?

72 m/z − 43 m/z = 29 m/z

19. What group likely left to form the base peak? (Look at Table 1 if you need a clue)

$H_3C-CH_2^+$

20. What fragment ion is represented by the base peak?

$H_3C-\overset{+}{CH}-CH_3$

21. Using similar strategies, assign possible fragment structures to the following *m/z* ratios:

a. 57 $H_3C-\overset{CH_3}{\overset{|}{C}H}-\overset{+}{C}H_2$ $^{+}CH_3$

b. 42 $\overset{CH_3}{\overset{|}{C}H}-\overset{+}{C}H_2$

c. 27 $\overset{+}{C}H-CH_2$

d. 29 CH_3-CH_2

22. Even though both 2-butanone (Figure 1) and 2-methylbutane (Figure 2) have the same molar mass, why do they have different mass spectra?

They fragment differently based on their original structures.

23. If you had an unknown compound, how could you determine from a mass spectrum if it was 2-butanone or 2-methylbutane?

Look at the ~~splitt~~ fragmentation pattern and look for masses of potential fragments

Consider this...

Table 2 *Isotope Abundances for Selected Elements*

Carbon	Hydrogen	Oxygen	Nitrogen	Chlorine	Bromine
C-12: 98.90%	H-1: 99.99%	O-16: 99.76%	N-14: 99.63%	Cl-35: 75.77%	Br-79: 50.69%
C-13: 1.10%	H-2: 0.015%	O-18: 0.20%	N-15: 0.37%	Cl-37: 24.23%	Br-81: 49.31%

Key Questions

24. Examine Table 2, and determine which elements have more than one isotope with significant (greater than 1%) abundance.

carbon, chlorine, bromine

25. Determine the masses of the following compounds to the nearest mass unit (whole number *m/z*). Assume the isotopes are the primary isotope unless specified.

a. CH_3-CH_2-Cl 64 m/z

b. $^{13}CH_3$-CH_2-Cl 65 m/z

c. CH_3-CH_2-^{37}Cl 66 m/z

26. Assume that a sample had all of these compounds in it at the appropriate abundances. Sketch the mass spectrum in the range of 60-70 mass units and label the axis.

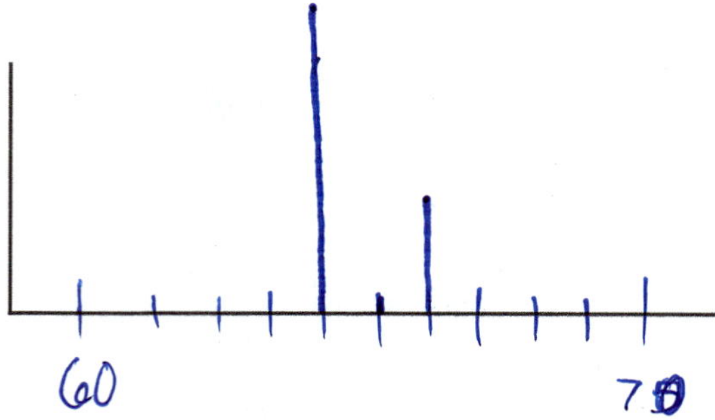

27. Sketch the mass spectrum near the parent ion for CH_3-CH_2-Br. Compare with your group.

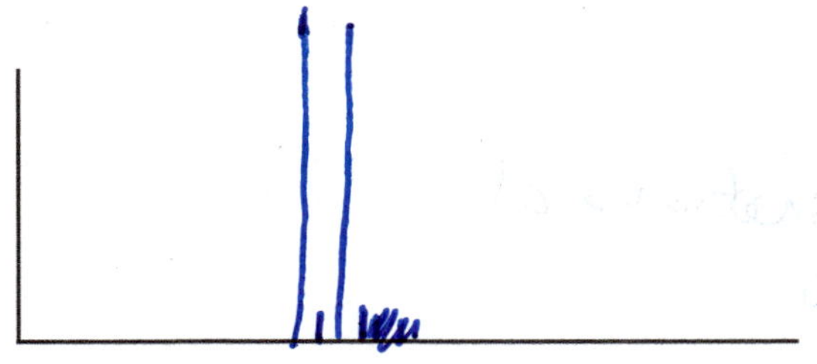

CH_3CH_2Br : 108

$^{13}CH_3CH_2Br$: 109

$CH_3CH_2{}^{81}Br$: 110

$^{13}CH_3CH_2{}^{81}Br$: 111

28. Compare your sketches in questions 25 and 26 with another group. Resolve any cifferences.

Consider this…

Figure 3 ***Relative intensities of ions for various ratios of chlorine and bromine***

	Br	Br_2	Br_3	Cl	Cl_2	Cl_3
M+	100	51	34	100	100	100
M + 2	94	100	100	31	65	95
M + 4		47	97		10	31
M + 6			31			3

29. Compare your sketches in Q26 and Q27 to the one chlorine and one bromine graphs in Figure 3. Explain why the mass spectra would have these patterns.

isotopes
and different combinations of
#s of isotopes

30. What is the ratio of the peak heights of the M+ : M + 2: M + 4 peaks for two chlorine?

100 : 65 : 10

31. What isotopes are responsible for each of the peaks in the two chlorine spectra?

~~35 Cl and 37 Cl~~

35 : 35 M+ 35 : 37 M+2 37 : 37 M+4

32. Confirm or refute: If a fragment has two bromines, its mass spectrum will have the pattern of peaks shown in Figure 3.

confirm

Consider this...

Figure 4 ***Mass spectrum of 1,4 dichlorobenzene***
(Used with permission of National Institute of Standards and Technology)

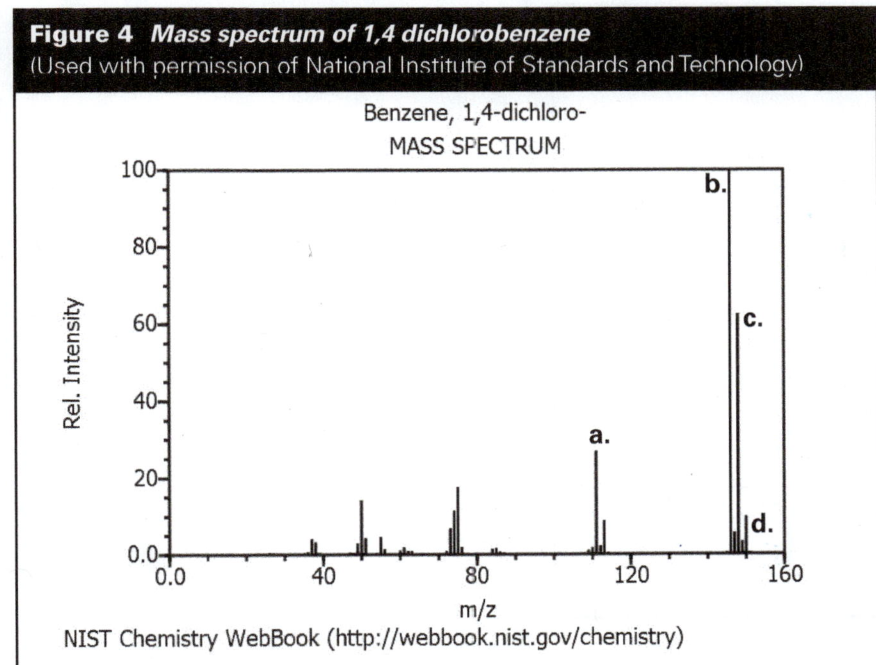

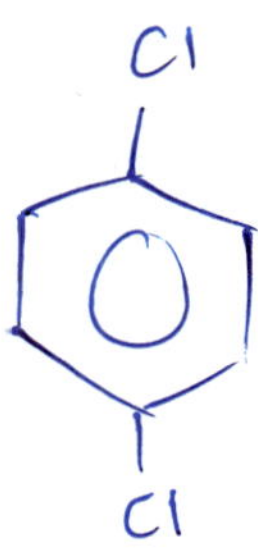

Key Questions

33. Examine the mass spectrum for dichlorobenzene and determine the fragments responsible for peaks a-d.

a.

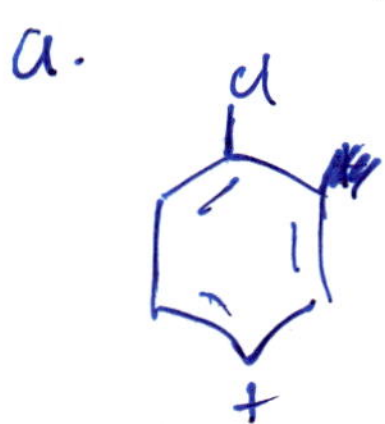

b.

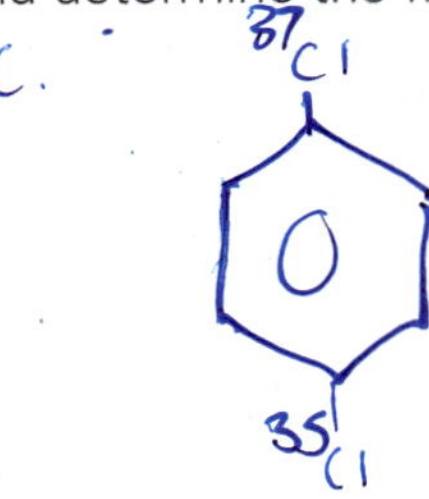

c. ^{37}Cl / ^{35}Cl

d. ^{37}Cl / ^{37}Cl

34. Discuss in your group and explain how you can determine if a compound contains either chlorine or bromine?

splitting patterns

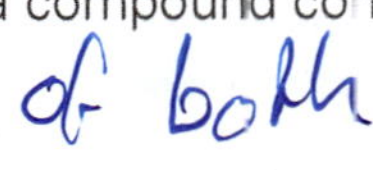

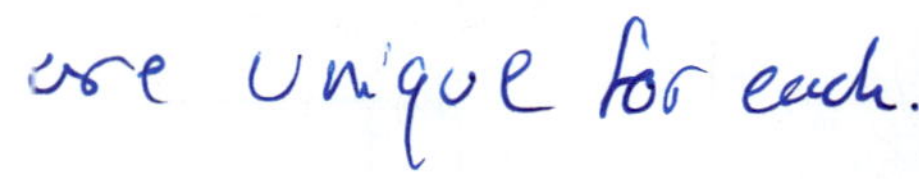

35. Examine the dichlorobenzene spectrum again. What peak is indicative of a benzene ring in the spectrum? How do you know?

76 m/z
molecular weight of (benzene) -2H

Consider this...

Figure 5 ***Mass Spectra of n-alkanes***
(Used with permission of National Institute of Standards and Technology)

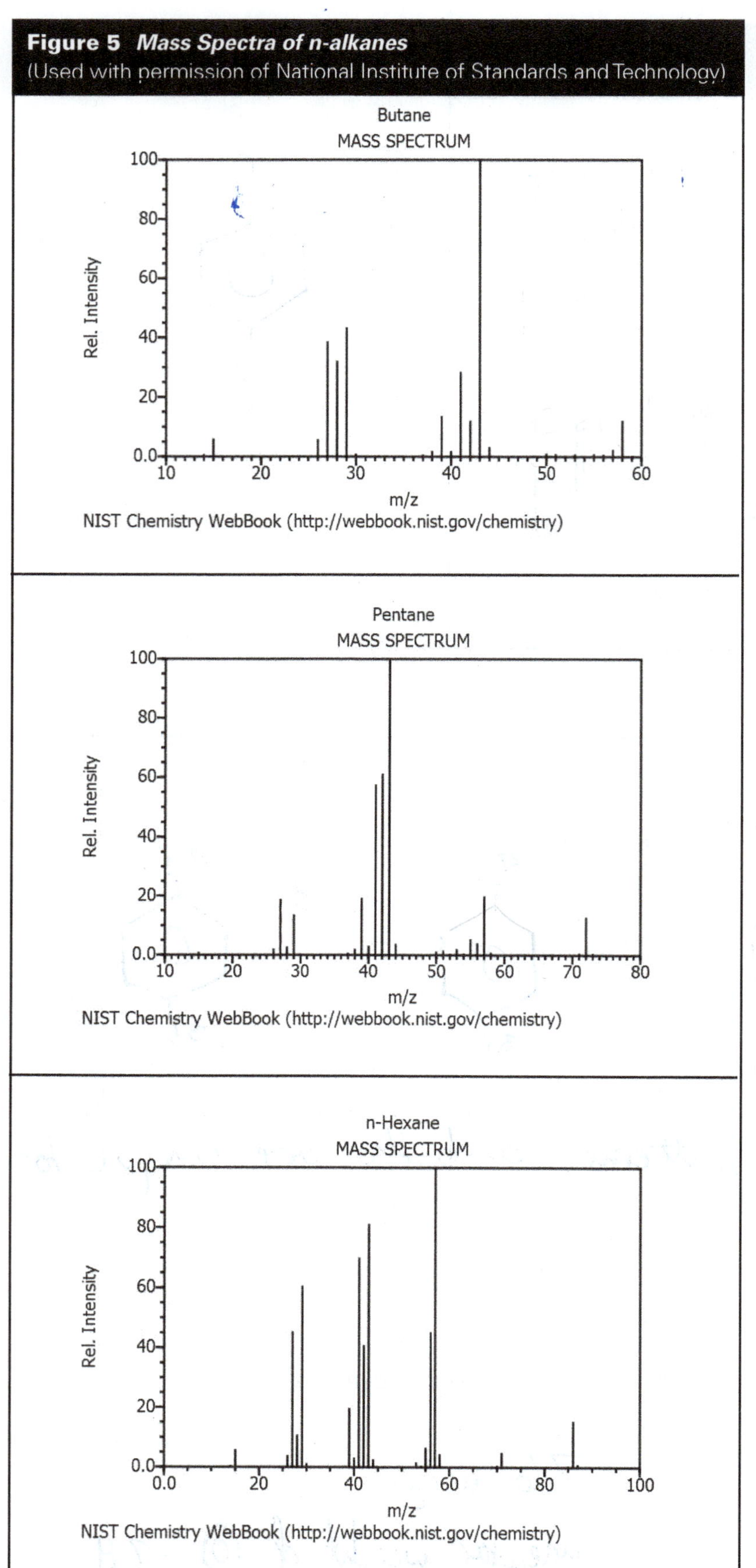

36. Examine the mass spectra of n-butane, n-pentane, and n-hexane. What patterns would you look for to identify an n-alkane?

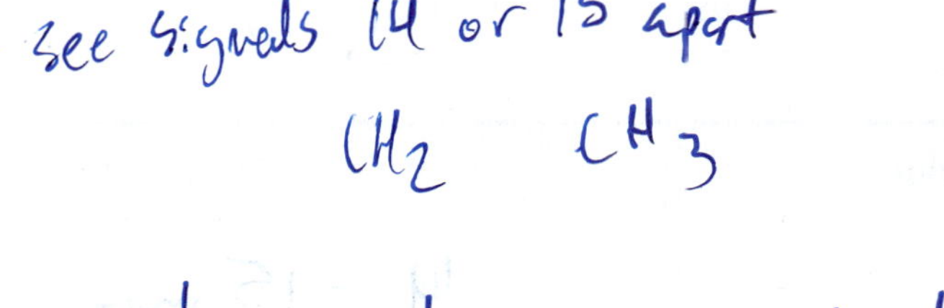

37. Based on your answer to Q31, which of the following would you expect to be an n-alkane? Explain your reasoning. (Used with permission of National Institute of Standards and Technology)

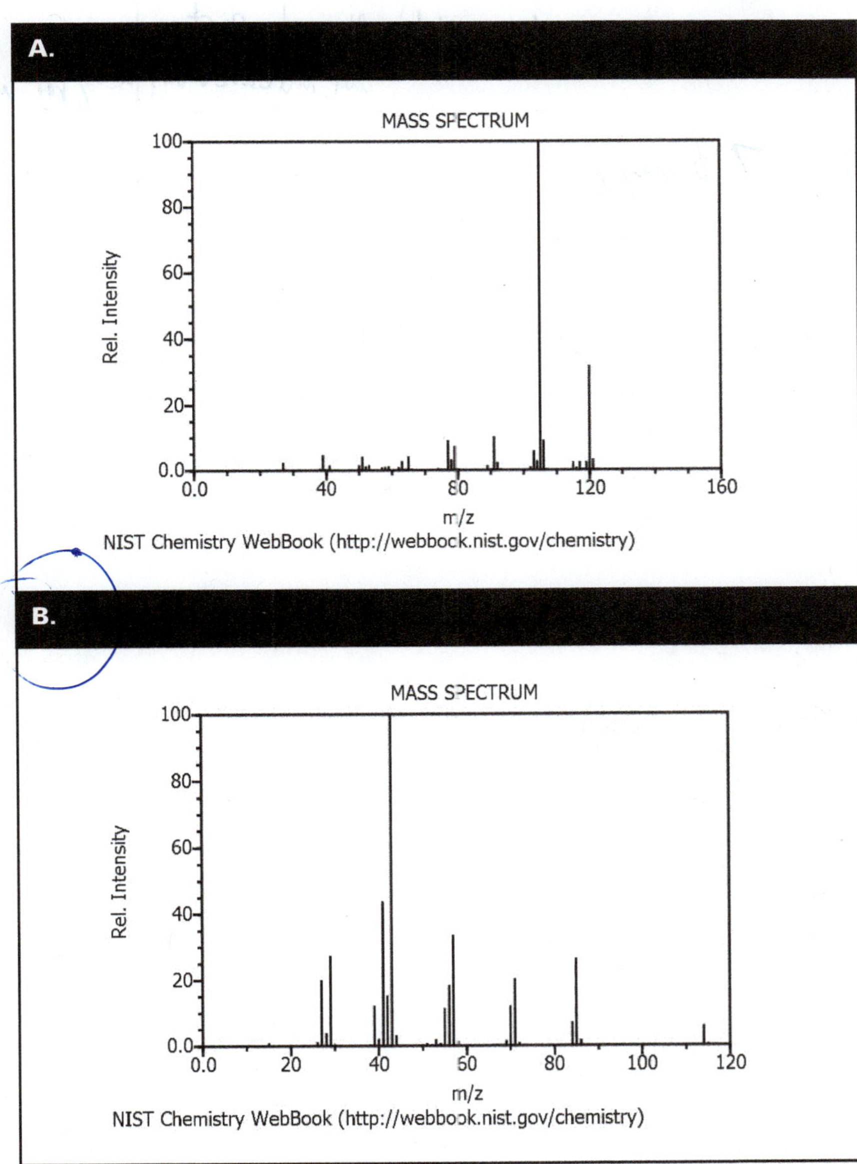

38. For each of the functional groups in the table below describe the pattern or peaks you would find in a mass spectrum to identify the group.

Functional Group	Mass spectrum pattern
Alkanes	14 – 15 m/z
Chlorine	← ~~[illegible]~~ 70-75% ← 25-30%
Bromine	2 apart for 1 (doublet), then 1 apart and increase multiplicity for more
Benzene Ring	78 m/z

39. List at least 3 ways you could use mass spectra to aid in identifying organic compounds.

40. How did this activity help you to identify patterns in a spectrum?

Application

41. Examine the following spectrum of chlorobenzene. (Used with permission of National Institute of Standards and Technology)

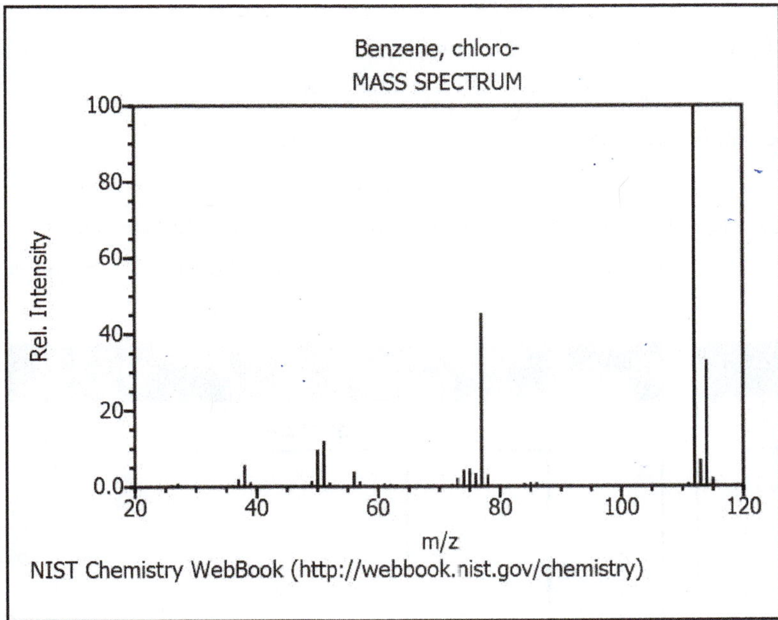

A. What is the mass of the base peak in this spectrum?

B. Identify the fragments that produce the three largest peaks.

42. Examine the two mass spectra below. One is 1-bromo-4-ethylbenzene and one is 2-bromoethyl benzene. Determine which is which and explain your reasoning. (Used with permission of National Institute of Standards and Technology)

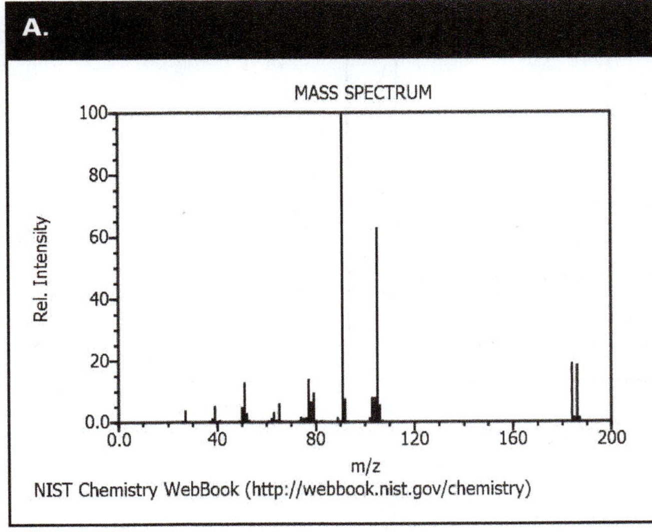

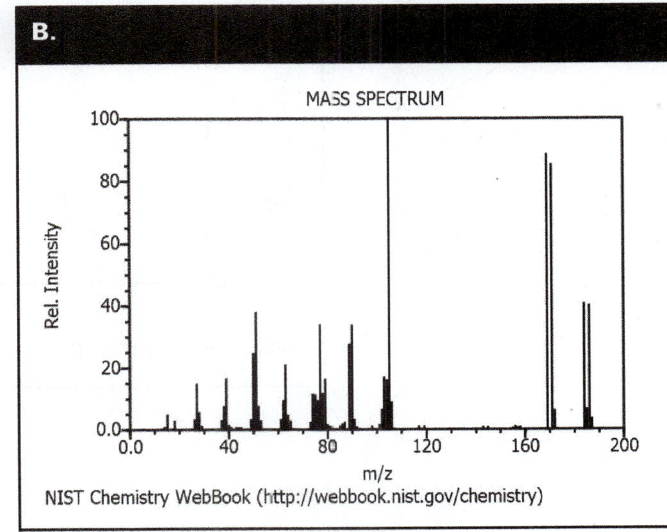

43. There are some additional "rules" that help in the interpretation of mass spectra.

a. Alcohols typically lose the hydrogen on the oxygen and often water (mass of 18) and have a very small parent ion.

b. Compounds that contain an odd number of nitrogen have an odd molecular mass.

c. Carbonyl compounds tend to cleave next to the C=O.

Based on these new rules determine which of the following spectra correspond to n-butyl amine, butanol, and diethylketone. Explain your reasoning. (Used with permission of National Institute of Standards and Technology)

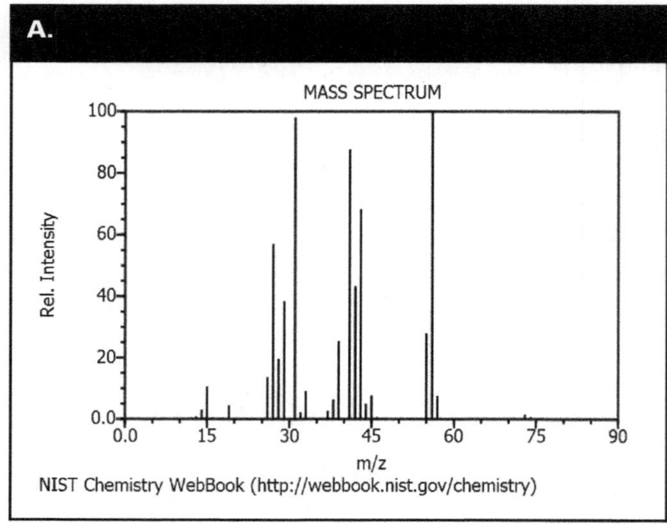

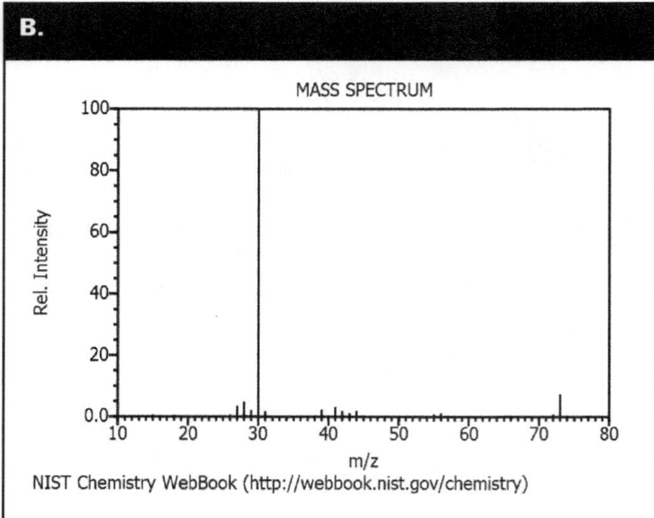

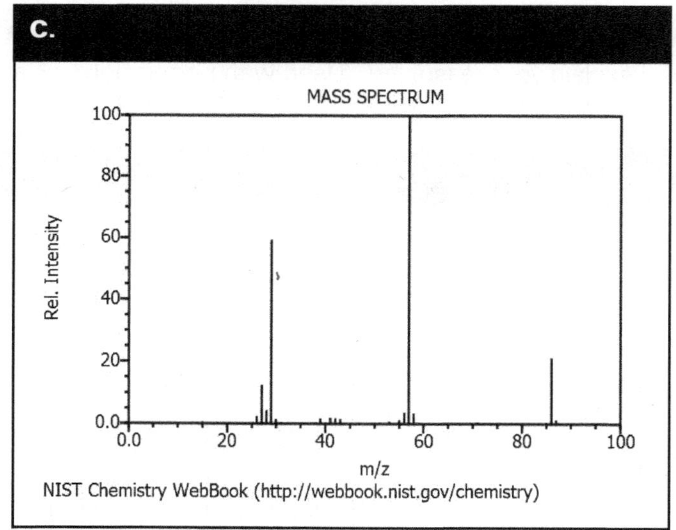

44. Examine each of the spectra below and provide as much information as possible about each compound. (Used with permission of National Institute of Standards and Technology)

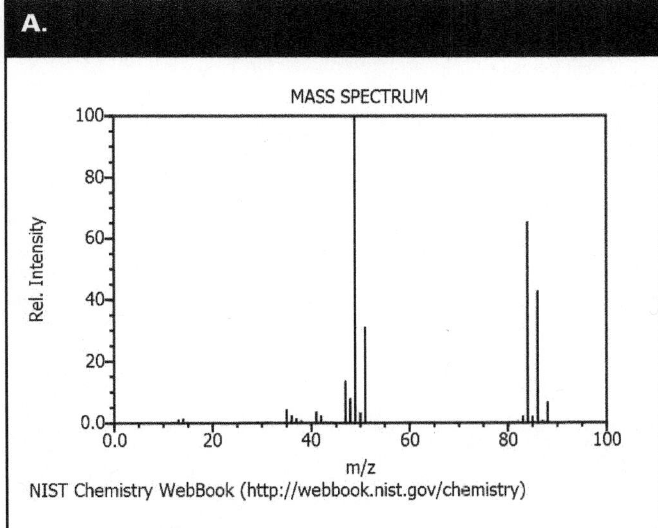

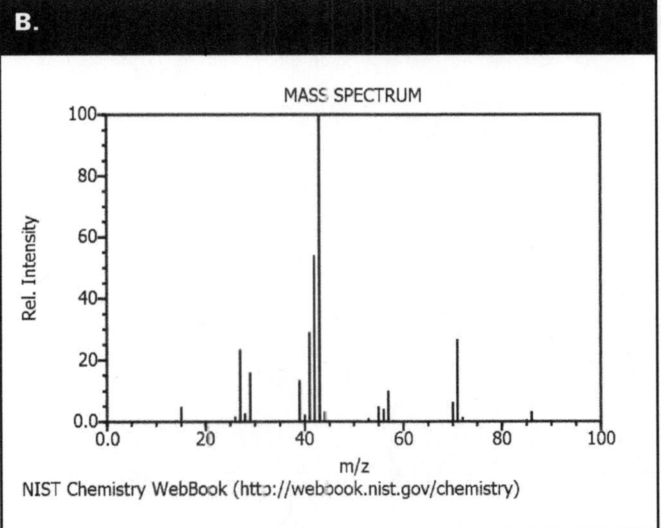

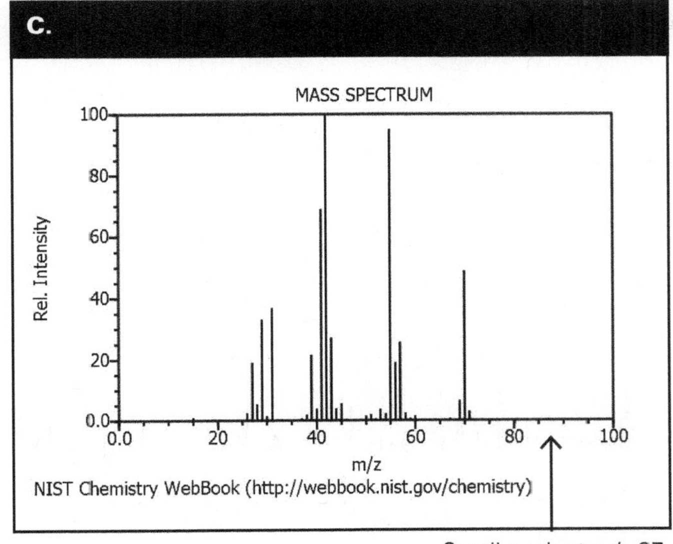

Small peak at m/z 87

Introduction to Chromatography

Learning Objectives

Students should be able to:

Content

- Relate the separation process to features of the chromatogram, such as retention time and resolution.
- Explain how the chemical interactions between a solute and both the stationary phase and the mobile phase impact retention.
- Predict the elution order for a set of compounds given the mobile and stationary phase composition.

Process

- Relate molecular interactions and macroscale processes (Critical Thinking).

Prior knowledge

- An understanding of intermolecular forces at the general chemistry level.

Further Reading

- Harris, D.C. 2010. *Quantitative Chemical Analysis,* 8th Edition, W.H. Freeman: USA, Section 22-2, p.542-3.
- Skoog, D.A., F.J. Holler, S.R. Crouch, 2007. *Principles of Instrumental Analysis,* 6th Edition, Thomson/ Brooks-Cole, 26A-B, pp.763-768.

Authors

Caryl Fish, Mary Walczak, Ruth Riter and Paul Jackson

Consider this...

In the experiment shown below in Figure 1, a glass column is packed with a solid material suspended in solvent. The solid material is called the stationary phase and the solvent is the mobile phase. Initially the same amounts of solutes A and B are added to the top of the column as shown in (a) below. As time progresses as shown in (b)-(e) solvent is allowed to flow through the column.

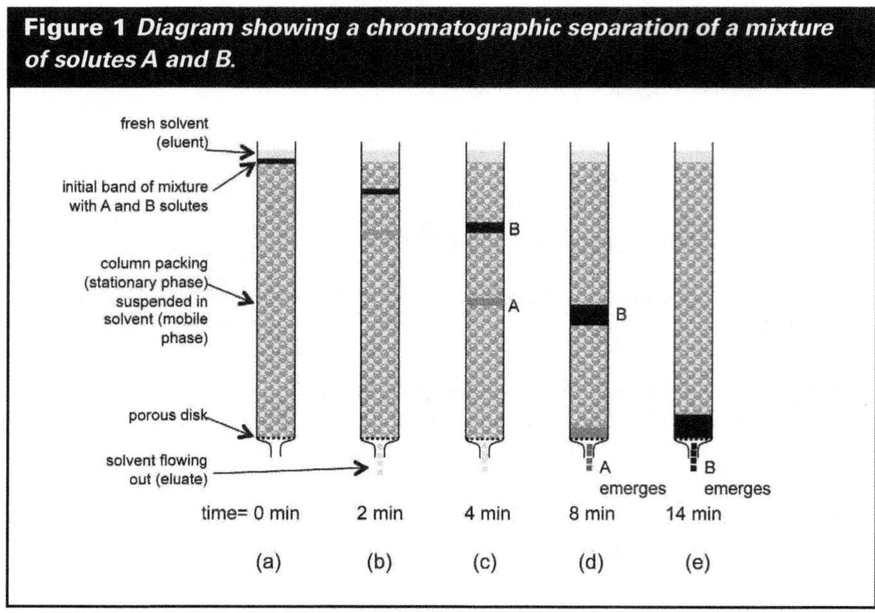

Figure 1 *Diagram showing a chromatographic separation of a mixture of solutes A and B.*

Key Questions

1. In Figure 1a, solutes A and B are mixed together in the initial band. As individuals, list differences observed as time progresses from (a)-(d). Compare your lists within the group. Develop a comprehensive list for the group.

- Solutes have diferent solubilities in the mobile phase
- A emerges from column first.

> The **retention time** (t_r) for a solute is the time needed after the mixture is placed on the column until that component emerges from the column. The retention factor (k') is the ratio of the time a solute spends in the stationary phase to the time spent in the mobile phase. Retention time for an unretained solute is the void time (t_m).

2. Using the elution times from Figure 1, what are the retention times for solutes A and B?

A: 8 min.
B: 14 min.

Consider this...

Figure 2 *A chromatogram.*

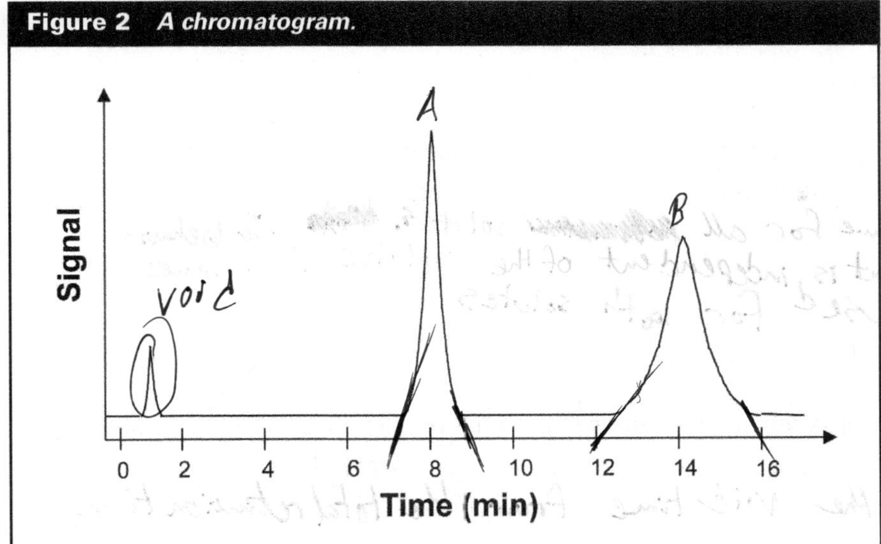

The plot of the signal detected at the outflow of the column in Figure 1 during the elution of the solute mixture A and B. The first peak represents the time it takes for an unretained substance to pass through the column; it has no interaction with the stationary phase.

Key Questions

3. Which peak on the chromatogram corresponds to solute A and which peak corresponds to solute B? Label each peak as either A or B. Does your group agree with these labels?

yeah.

4. The retention time determination should be consistent from person to person. From the chromatogram what part of the peak should be used to give the most consistent retention times? Explain your reasoning to your group.

The center/middle/peak of the peak.

It should give an approximately average time.

5. Looking at the chromatogram, determine the void time for this column.

1 min.

6. The void time is the same for all the solutes in this chromatogram. Discuss with your group why the void time is equal to the time the solutes spend in the mobile phase. Record your agreed upon explanation.

It is the same for all solutes. The behavior of the solvent is independent of the solutes. The same solvent is used for both solutes

7. If the retention time of a solute is the time the solute spends in both the mobile phase and the stationary phase, how could you determine the time the solute spends in only the stationary phase?

Subtract the void time from the total retention time.

8. Write an equation to show how the retention factors for each solute are determined from the chromatogram in Figure 2 and calculate the retention factors for solutes A and B. Note that the retention factor is a unitless number.

$k' = \frac{S}{M}$ $k'_A = \frac{7 \text{min}}{1 \text{min}} = 7$ $k_B' = \frac{13 \text{min}}{1 \text{min}}.$

9. What is the relationship between the broadness of the peaks A and B in Figure 2 and their bandwidths shown in Figure 1?

B elutes slower than A in the solvent used. It is detected over a longer time.

10. Describe the relationship between the retention factor and characteristics of the peaks on the chromatogram such as retention time and peak broadness.

The retention factor affects how strongly the solute binds to the column. A higher retention factor means a higher retention time because the sample sticks to the column longer. This also likely means broader peaks because it'll take longer for the solutes to completely leave the column.

In chromatography the term resolution (R_s) is used to express the quality of a separation between two peaks.

11. Look at Figure 2 and discuss with your group how changes in retention time and the broadness of peaks A and B would influence R_s. Write a summary of your group consensus below.

Closer retention times for both A and B would reduce resolution because they would elute at similar times without as much separation.

12. Resolution can be computed by dividing the difference in retention times between the two peaks by their average width along the baseline. f a value of R_s=1.50 represents baseline resolution between two peaks in time units, will the R_s value for the separation shown in Figure 2 be greater than or less than 1.50? Briefly explain your answer, and then verify it with a quick calculation by estimating these values.

~~14-8~~ $\frac{6}{2.5}$ = 2.4

14-8 = 6

A: $\frac{6}{2}$ = 3 B: $\frac{6}{4}$ = 1.5

Should be greater than 1.50 because there looks to be enough separation between the peaks.

13. Based on your understanding of resolution, speculate as to how a chemist might improve the resolution of two chromatographic peaks.

- longer column
- change solvent polarity
- change amount of mobile phase (increase)

Consider this…

Figure 3

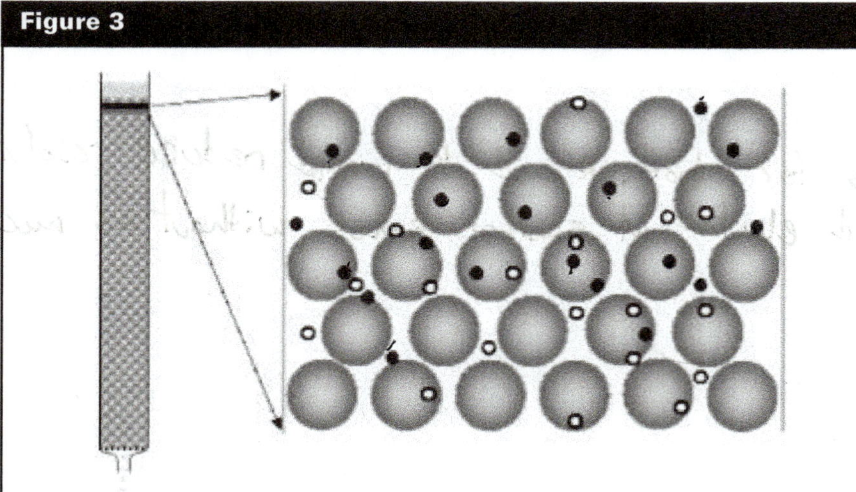

This is a snapshot microscopic view of individual solute molecules A (open circles) and B (black circles) in the vicinity of a solid stationary phase porous spherical particle (silica, SiO_2 with a large number of surface Si-OH groups) in a column. Those solute molecules touching a stationary phase particle should be considered in the stationary phase. The mobile phase (hexane, C_6H_{14}) in which the solute is dissolved is not shown.

Key Questions

14. Examine the distribution of solutes A and B between the mobile phase and the stationary phase. Which solute has more molecules in the mobile phase?

A ; the A particles seem to be moving down the column faster than B and don't seem to bind as much to the silica gel.

15. This distribution of the solute molecules is similar to a partitioning between the stationary phase and the mobile phase. We can define a distribution constant (K) as the ratio of the concentration of the solute in phase 2 to the concentration in phase 1. In chromatography the mobile phase is considered phase 1 and the stationary phase is phase 2. Write the distribution constant expressions for solute A (K_A) and B (K_B).

$K_A = \frac{13}{6}$ $K_B = \frac{17}{3}$ $\frac{s}{m}$

16. Using the number of solute molecules in each phase in Figure 3, quickly approximate the distribution constants for solutes A and B.

$K_A = \frac{13}{6}$ $K_B = \frac{17}{3}$

17. If the interaction between the mobile phase and the solute is stronger than the interaction between the stationary phase and the solute, which phase do solute molecules partition preferentially into, the mobile or stationary phase? One group member should explain his/her reasoning to the group.

Solute molecules will prefer partitioning into mobile phase. Likely due to higher solubility.

Keep in mind that Figure 3 is a snapshot as if all movement magically stopped in the column. Obviously, this does not occur during the separation process. Remember that this is a dynamic process and solute molecules are constantly moving between the mobile phase and the stationary phase. In addition, the process of separation is not an equilibrium process.

18. The solute will move through the column only if it is in which phase?

mobile

19. If a solute molecule partitions preferentially into the mobile phase, would you expect it to have a relatively long or short retention time? Do the relative values of K_A and K_B in KQ 16 support your conclusion? Explain.

Short Retention time. Yes, they do. B seems to have a stronger stationary phase interaction, which explains its higher retention time.

20. Re-read the description of the mobile phase and stationary phase in Figure 3. Describe the mobile and stationary phases as polar or non-polar.

Stationary phase is polar because of the large # of Si-OH on the silica gel particles. The mobile phase is nonpolar because it's just a hydrocarbon.

21. Discuss with your group whether solute A or B is the most polar, given that Solute B is retained longer on the column. Once you have reached a consensus, describe your group's reasoning below.

B is more polar than A because it binds more effectively to the stationary phase, which is also polar. Thus, it has a stronger interaction with the stationary phase and is retained longer.

22. Discuss and explain in complete grammatically correct sentences your group's consensus of why Solute A elutes before solute B.

Solute A will elute first because it is less polar than B, so it has a weaker binding interaction with the stationary phase than solute B.

23. How confident are you in predicting the elution order of compounds in a separation? If you are not confident, what additional questions do you have?

Decently. To what extent does each variation of a substituent on a compound (ie. alkyl, nitro, amine, ether, hydroxy, etc.) contribute to its overall polarity and column binding.

24. What is one way this activity improved your ability to visualize the microscopic separation processes embedded in the appearance of a chromatogram?

Concentration of a band on a band on a column correlates with its peak signal and width.

Applications

25. In the experiment illustrated in Figures 1-3, which of these variables could change the elution order? Explain your reasoning.

Mobile phase composition	Length of column	Solvent flow rate
Solute composition	Stationary phase composition	

26. Suppose you wish to decrease the amount of time for all the peaks to elute but continue to completely separate the compounds in Figure 2. Which of the following variables would affect this change? Explain your reasoning.

Mobile phase composition	Length of column	Solvent flow rate
Solute composition	Stationary phase composition	

27. Which of the variables you identified in Q26 would be easiest to change and give the desired effect? Explain your reasoning.

28. Suppose you repeat the experiment in Figures 1 and 2 with the same mobile phase and stationary phase, but using methanol and cyclohexane as solutes. Which solute would elute first? Explain your reasoning.

29. Stationary phases come in different varieties. Which of the two analytes listed for each stationary phase has the greater retention time (or is retained more)? Explain your answer.

Stationary Phase	Retention mechanism	Analytes
Polar Solid	adsorption	H_2O and benzene
Nonpolar Liquid	solubility	H_2O and benzene
Ionic	electrostatic interaction	Na^+ and Mg^{2+}
Porous solid	pore penetration	ethylene and polyethylene

Band Broadening Effects in Chromatography

Learning Objectives

Students should be able to:

Content

- Explain how band broadening (i.e., band dispersion) is related to linear flow rate.
- Discuss how the three components in the van Deemter equation affect bandwidth.

Process

- Given a plot of data, identify the mathematical relationship (Critical thinking).
- Sketch graphs of known mathematical relationships (Information processing).

Prior knowledge

- Identify polar and nonpolar molecules and solvents.
- Intermolecular forces.
- Introduction to chromatography and chromatograms.

Further Reading

- D.C. Harris, *Quantitative Chemical Analysis,* 7th Edition, 2007 W.H. Freeman: USA, Section 23-2, p.516-521.
- D.A. Skoog, F.J. Holler, S.R. Crouch, *Principles of Instrumental Analysis,* 6th Edition, 2007, Thomson/ Brooks-Cole, 26A-B, pp.771-775.
- *The van Deemter Equation: A Three-Act Play* by Christa Colyer, Department of Chemistry, Wake Forest University, Winston-Salem, North Carolina. http://ublib.buffalo.edu/libraries/projects/cases/vandeemter/vandeemter.html
- S. J. Hawkes. "Modernization of the van Deemter Equation for Chromatographic Zone Dispersion." *J. Chem. Educ.* 1983, 60(5), pp.393-398.
- Patel, et al. *Anal. Chem.* 2004, 76, pp.5777-5786.
- Kirkland, J.J. *American Laboratory* April 2007, 18-21; Li & Carr. *Anal. Chem.* 1997, 69, pp.2193-2201.
- Mattice, John. "If You Were a Molecule in a Chromatography Column, What Would You See?" J. Chem. Educ. 2008, 85(7), pp.925-928.

Authors

Caryl Fish, Paul Jackson, Mary Walczak, Ruth Riter (Shepherd)

Consider this...

Figure 1

The left illustration (A) shows a glass chromatography column packed with porous, spherical particles surrounded by solvent. Two solute bands are also shown, and the mobile phase is flowing from top to bottom of the column. The center drawing (B) is a microscopic view (not drawn to scale) of 14 particles in this column. Five (1-5) solute molecules are shown for simplicity. All the space between the stationary phase particles is filled with mobile phase which is not shown in the figure. On the right (C), a picture of a single, porous, spherical particle obtained with a scanning electron microscope (SEM) is shown.

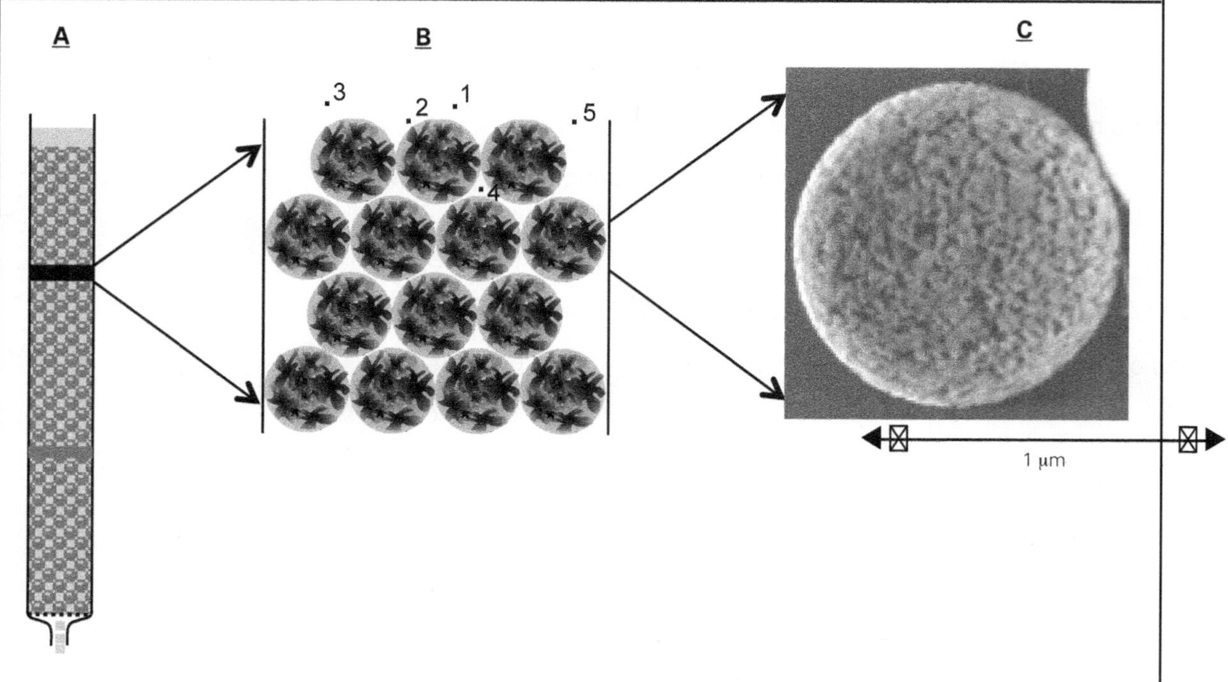

Key Questions

Multiple Paths

1. Consider solute molecule 1 in Figure 1 (B). Each group member should independently draw a path for the molecule to move through this section of the column.

2. Compare your results. Which path is shortest? Which path is longest?

[1]Jacoby, Mitch. CE&N. 2008, 86(17), pp.17-23.

3. Considering the other molecules (2-5), draw possible paths for these molecules as they move through the figure. What effect would a large number of different solute paths have on band broadening?

4. Draw chromatograms for a solute eluting from a column with
A) a small number of multiple paths? B) many multiple paths?

5. As the mobile phase flow rate increases, will the *number* of pathways change as the mobile phase flow rate increases? Based on this, sketch in a graph the relationship between bandwidth and mobile phase linear flow rate, u_x.

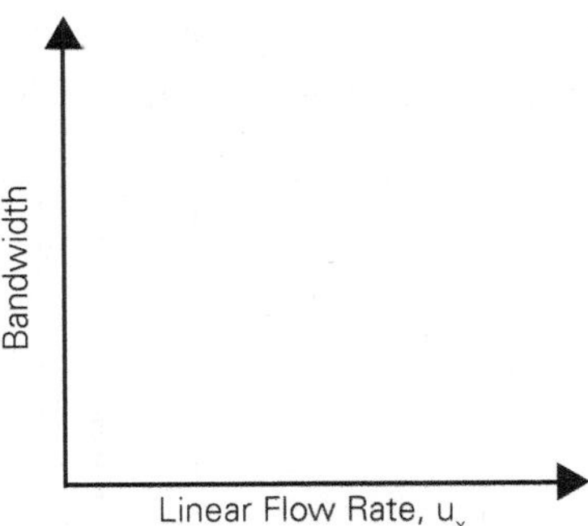

6. Write a proportionality between bandwidth and linear flow rate u_x. (*Hint:* bandwidth may be directly proportional, inversely proportional, or constant as u_x increases.)

Longitudinal Diffusion

7. Consider solute molecule 4 in Figure 1 (B). What is the most likely direction that this molecule will travel through the column while the mobile phase is flowing?

8. If we stop the mobile phase from flowing, solute molecule 4 will continue to move in the column due to diffusion. Draw arrows around this molecule to indicate the directions it might move.

9. What effect will diffusion of the solute molecules have on the broadness of the peaks eluting from the column?

10. Band broadening due to diffusion depends on flow rate. Will band broadening due to diffusion be more pronounced at slow or fast flow rates?

11. The relationship between the broadness of the peaks due to diffusion and mobile phase flow rate is represented by one of the following three graphs.

A) Choose which of the three models below represents this relationship and explain your rationale.

B) Write a proportionality between bandwidth due to diffusion and linear flow rate u_x.

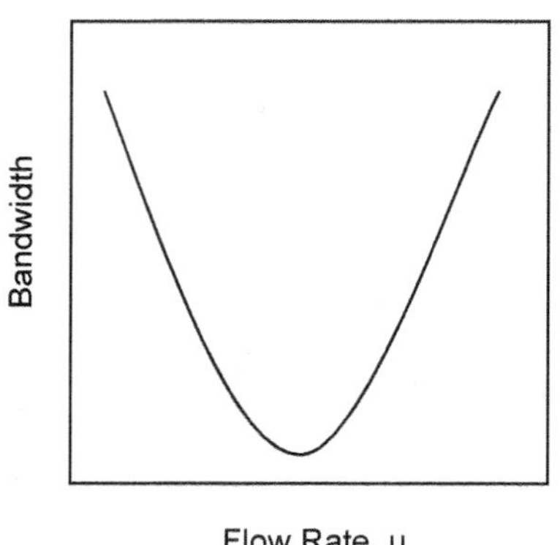

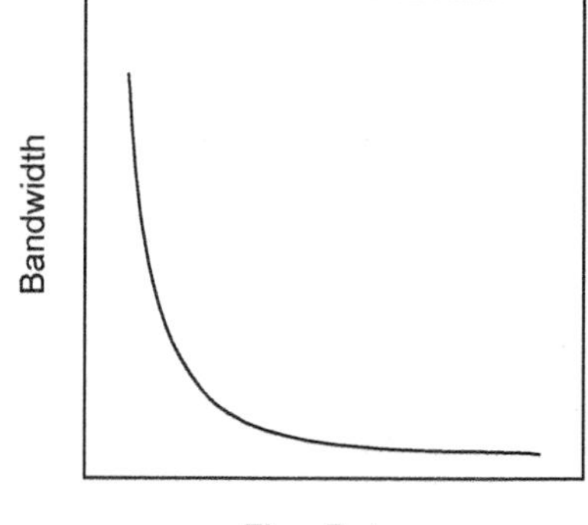

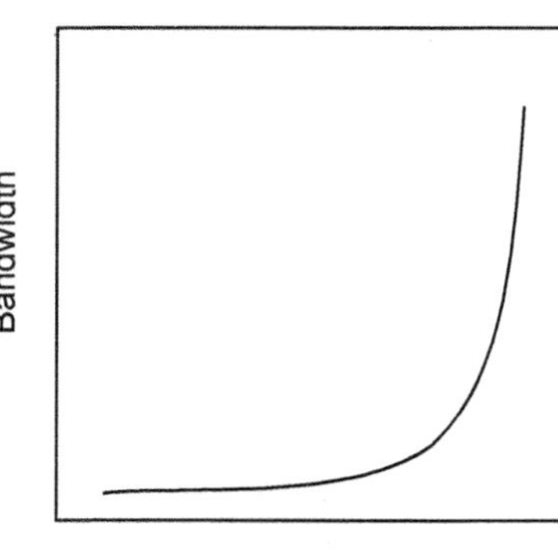

Mass Transfer

Chromatographic separation is a dynamic differential migration process. Solute molecules "distributed" or "dissolved" in the stationary-phase particles transfer to the mobile phase and vice versa continually throughout the column.

12. In Q1 and Q3 above, did your paths for molecules 1-5 include the molecules interacting with the porous stationary-phase particles? Yes ___ / No___. Explain why the answer should be yes. If your answer is no, each group member choose one solute molecule in Figure 1 (B) and independently sketch a new path where the retention time is longer because the solute molecules interact with the stationary-phase particles.

13. Compare your drawn paths and briefly describe a difference and a similarity.

14. As the solute molecules move through the column in the mobile phase, **(a)** a portion of solute molecules spend little time interacting with the stationary phase and **(b)** a portion of solute molecules spend long periods of time in the stationary phase. Compare the progress of the **(a)** molecules down the column to the progress of the **(b)** molecules. Would band broadening increase or decrease due to the mass transfer of solute molecules to the stationary phase?

15. Now predict the progress of the **(a)** molecules if the flow rate of the mobile phase is increased.

16. Would you expect band broadening due to mass transfer to increase or decrease when the flow rate of the mobile phase increases? Explain your reasoning.

17. Assume that the relationship between bandwidth due to mass transfer and flow rate is linear. Write a proportionality between bandwidth and linear flow rate, u_x, and sketch a plot of this relationship.

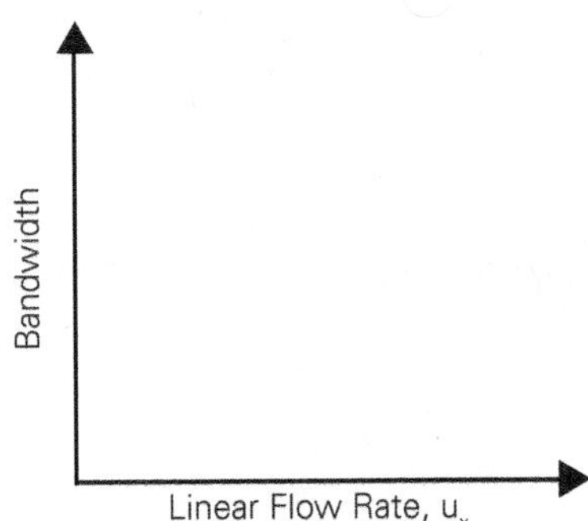

18. The moving of solute molecules from the mobile phase to the stationary phase and vice versa through the column is called mass transfer. Explain why and compare your group's answer with at least one other group in your class.

Consider this...

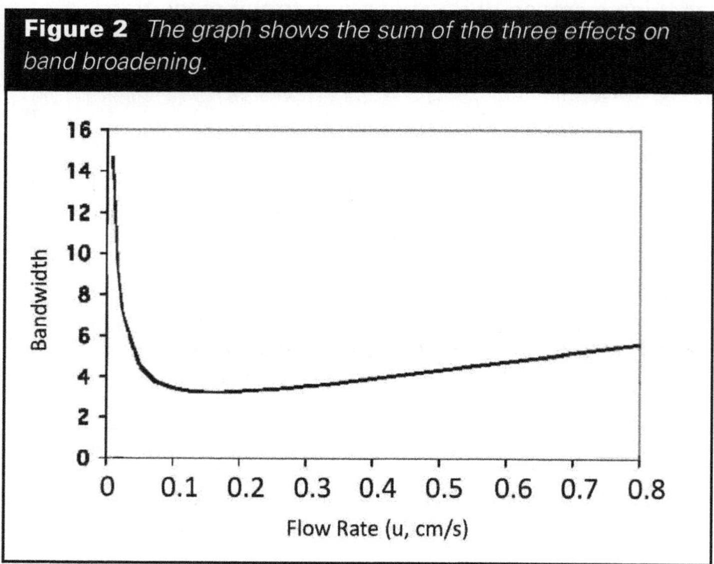

Figure 2 *The graph shows the sum of the three effects on band broadening.*

Key Questions

19. At flow rates < 0.05 cm/s, which of the three processes (multiple paths, mass transfer, or longitudinal diffusion) dominates band broadening?

20. At high flow rates (> 0.5 cm/s), which of the three processes is dominant?

21. Based on the three processes that lead to band broadening shown in Figure 2, what flow rate would you use to minimize band broadening?

22. Sketch on the plot above each of the three band broadening effects.

23. Would the bandwidth ever reach zero? Explain why or why not.

The relationship between band broadening and flow rate is known as the van Deemter equation where band broadening is defined in terms of "height equivalent to a theoretical plate," H:

$$H = A + (B/u) + Cu$$

The smaller the plate height, the narrower the bandwidth.

This equation is a useful qualitative model to describe how the three microscopic processes affect column dispersion.

24. Identify the three additive terms in the van Deemter equation using the relationships between the microscopic processes (multiple path, diffusion, and mass transfer) and mobile phase flow rate that you developed in Q6, Q11 and Q16.

25. After working through this activity and given the proportionality between an independent and dependent physical variables, how confident are you that you could sketch a graph

Applications

26. Predict how stationary-phase particle size would affect multiple paths band broadening mechanism and hence bandwidth.

27. In general, solute molecules in pores of the stationary-phase particles typically diffuse a distance equivalent to the particle diameter to leave the particle. Predict how stationary-phase particle size would affect mass transfer and hence bandwidth.

28. If the stationary-phase particle size in a column is decreased, does bandwidth increase, decrease, or stay the same? Explain your reasoning.

29. If the solute molecules and stationary phase are both nonpolar, how would increasing the polarity of a mobile phase affect bandwidth?

30. For gas chromatography, in an open tubular column where the stationary phase lines the walls of the column (as opposed to a packed column), A=0 in the van Deemter equation. Explain why this is true.

31. Increasing temperature of a chromatographic separation allows the flow rate of the mobile phase to be increased by a factor of 5 while maintaining comparable bandwidths. Explain how temperature affects mass transfer processes and hence bandwidth.

32. What impact (increase, decrease, or no change) does each of the following conditions have on the individual components of the van Deemter equation and consequently, band broadening?

	Multiple Paths	Diffusion	Mass Transfer
Increase temperature			
Longer column			
Using a gas mobile phase instead of liquid			
Smaller particle stationary phase			

33. Below are experimentally determ ned van Deemter plots of column eff ciency, *H*, vs. flow rate. *H* is a quantitative measurement of band broadening. The left plot is for a liquid chromatography application and the right is for gas chromatography. Compare and contrast these two plots in terms of the three band broadening mechanisms presented in this activity. How are they similar? How do they differ? Justify your answers.[2]

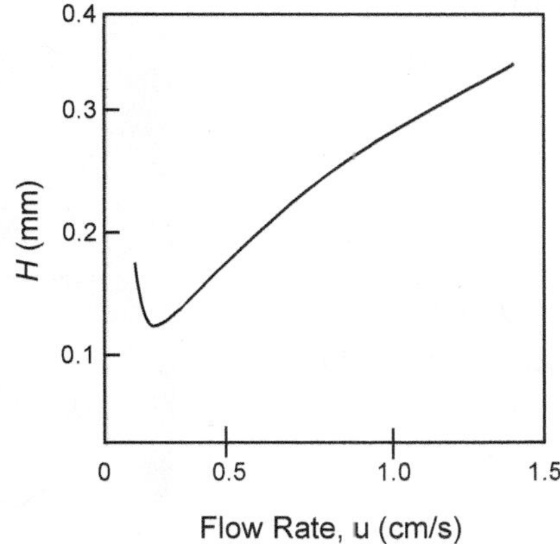

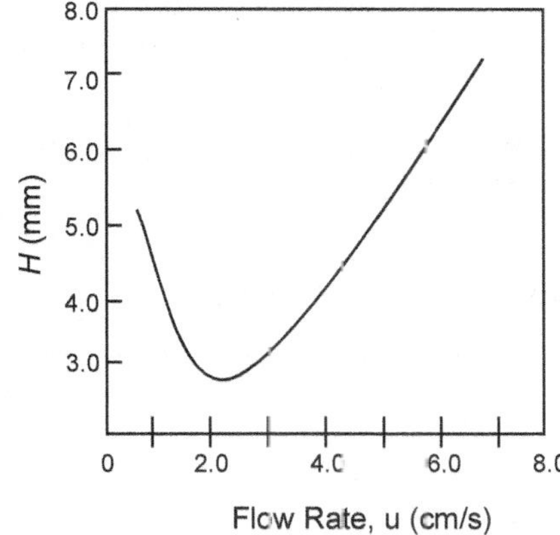

34. Figure 3 shows Van Deemter plots for a solute molecule using different column inner diameters (i.d.).

A) Predict whether decreasing the column inner diameters increase or decrease bandwidth.

B) Predict which van Deemter equation coefficient (A, B, or C) has the greatest effect on increasing or decreasing bandwidth as a function of i.d. and justify your answer.

Figure 3 *Van Deemter plots for hydroquinone using different column inner diameters (i.d. in μm). The data was obtained from liquid chromatography experiments using fused-silica capillary columns packed with 1.0-μm particles.*[3]

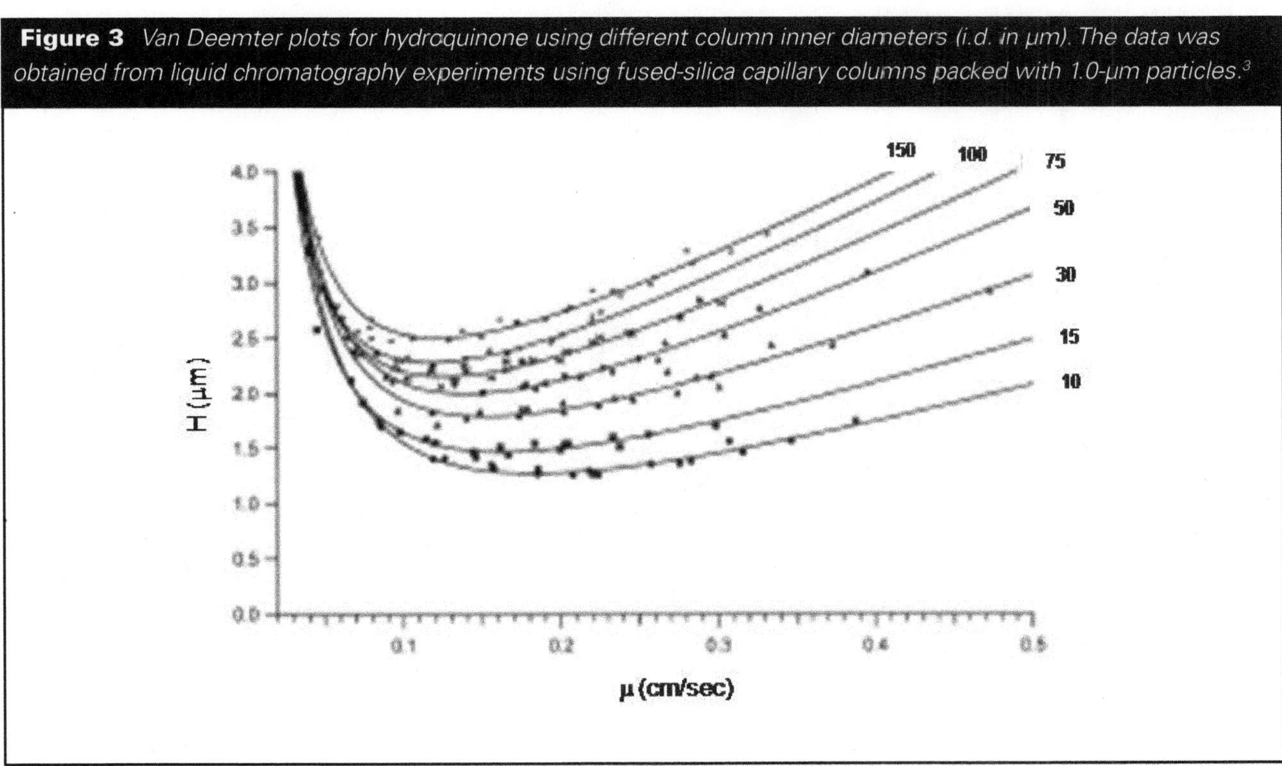

[2] Plots are adapted from Figure 30-13 in Skoog, D. A. S. M. West, F. J. Holler, and S. R. Crouch. *Fundamentals of Analytical Chemistry,* 8th ed., Belmont, Ca:- Brooks/Cole, 2004.

[3] Patel, et al. Anal. Chem. 2004, 76, pp. 5777-5786

35. Figure 4 shows scanning electron microscope (SEM) images of extruded sections of packing bed for two capillary columns of different diameters, **a)** 750 (bottom image) and **b)** 30-μm-i.d. Both columns are packed with the same stationary phase, spherical particles with 1-μm diameter.

A) When the columns were prepared, the figure shows that the column with the larger diameter has more packing irregularities. Explain this observation.

B) Predict what affect this should have on band broadening and discuss your prediction using the van Deemter terms.

C) Does this figure support your explanations in application question 33? Explain why or why not and make any changes in your answers in light of this figure.

Figure 4 *SEM images of sections of packed columns for a) 750 and b) 30-μm-i.d. capillary columns.*[3]

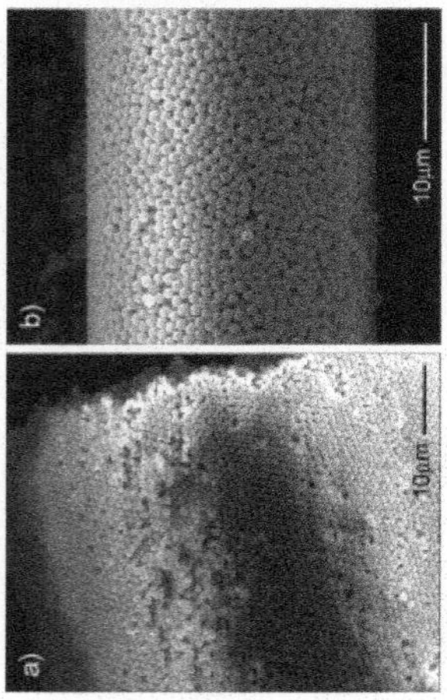

Gas Chromatography or HPLC, Which Do You Choose?

Learning Objectives

Students should be able to:

Content

- Describe the differences and similarities between Gas Chromatography (GC) and High Performance Liquid Chromatography (HPLC).
- Compare sample preparation for GC and HPLC.
- Compare resolution, detection limit, and analysis time for a particular chromatographic separation.

Process

- Interpreting schematic diagrams (Information Processing).
- Identifying similarities and differences (Critical Thinking).
- Integrating prior knowledge to choose appropriate instrument (Critical Thinking).

Prior knowledge

- Introduction to chromatography.
- Band Broadening.
- Gas chromatographic and HPLC instrument details.

Further Discussion

- Instrument components including advantages and disadvantages of different component types.
- Determining concentrations from chromatographic data.

Further Reading

- D.C. Harris, *Quantitative Chemical Analysis*, 7th Edition, 2007 W.H. Freeman: USA, Chapters 24-25, pp.528-583.
- Gratz, L.D. et al. *J. Hazardous Materials,* 74 (2000) 37-46.

Author

Caryl Fish

Consider this...

Figure 1 ***Diagram of a GC instrument***

Typical GC configuration: Helium as the carrier gas, injector heated to 250° C, 1 μl of sample is injected, the column is a 30-meter, open tubular column, the column oven can range from 60° C to 300° C, the detector is a flame ionization detector that responds to most organic compounds, the output is a computer system that can record the retention time and area of each peak, but does not give structural information about the analytes.

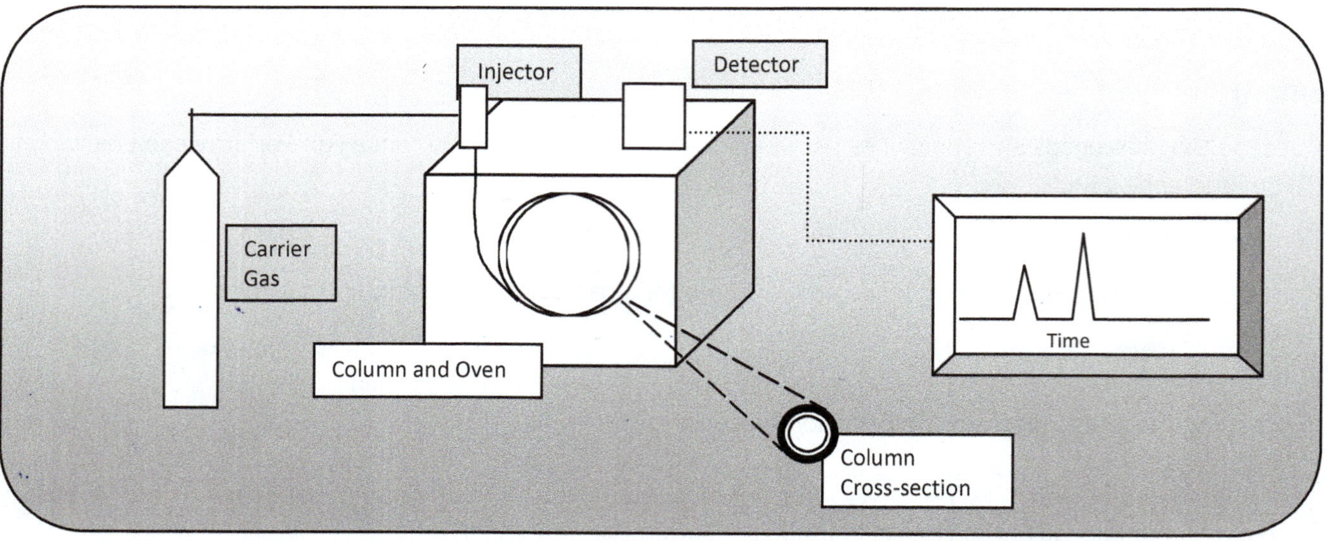

Figure 2 ***Diagram of an HPLC instrument***

Typical HPLC System: Solvent is a mixture of water and a polar organic solvent, the ratio of solvents can be changed, the pump flows at 1-2 μl/min, 10–25 μl of sample are injected, an analytical column is made of non-polar organic compounds bonded to 5–10 um particles packed into a 20 cm long column, the detector is a ultraviolet detector that records absorbance at a particular wavelength, the output is a computer system that can record the retention time and area of each peak, but does not give structural information about the analytes.

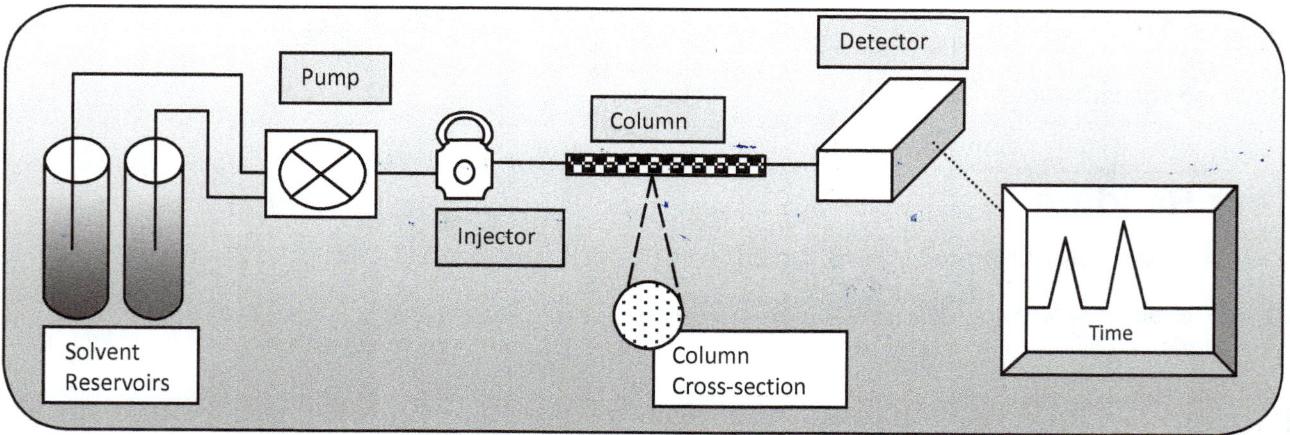

Key Questions

1. Determine three similarities and three differences between the GC and HPLC instruments. Compare your answers with your group and make a combined list.

Sims
- can record ret. time/peak area
- doesn't give structural information
- uses fractions of mLs for analyte

columns
detectors
output

Diffs
- UV detector on HPLC / flame ionization det. on GC.
- GC helium carrier gas / HPLC mix of solvents
- GC column is open / HPLC column made from non-polar organic compounds on 5-10 μm particles

GC narrower column, GC has oven, GC column is open, GC is gas mobile phase

2. Focus on the columns and examine the columns in the GC and HPLC. As a group, determine four differences.

- GC column is open, not packed
- GC mobile phase is a carrier gas, not a liquid solvent
- GC column is 30 m-long, HPLC column is 20 cm
- GC column is heated

3. Since the column is the heart of the separation process, describe how the differences in the GC and HPLC columns could impact the ability to separate compounds.

longer columns give better separation and eliminates band broadening → better separation

The GC column is open and unpacked, while the HPLC is packed. The GC uses helium as a carrier gas, and a compound mixture separates into constituents over a much longer distance in the 30-m column. The solvent mixture largely determines separation in HPLC in a much shorter 20-cm column.

4. What is the mobile phase in the GC? In the HPLC?

GC – Helium gas

HPLC – mixture of water and a polar organic solvent (liquid)

5. Based on these mobile phases, what characteristic(s) must an analyte possess to be able to use a GC for analysis? An HPLC for analysis?

GC – low melting points?, high volatility, thermally stable.

For HPLC – ~~reasonable~~ polar ~~and nonpolar solubility to be carried by the~~ solvent ~~mixture and interact~~ with ~~a nonpolar-coated column.~~

polar – soluble

Typically organic compounds analyzed by GC must be volatile (with boiling points less than 500° C) and thermally stable (do not decompose at high temperatures). For reverse phase HPLC (polar solvent, non-polar stationary phase), the analytes must be soluble in polar solvents.

Consider this...

Sample preparation for GC and HPLC can have four purposes: **1)** put the analytes into the appropriate solvent for the instrument; **2)** adjust the concentration to fit the concentration range of the instrument; **3)** remove contaminants that could interfere with the analysis; **4)** chemically react the analytes to change their properties so they are more amenable to separation and detection (derivatization). For example, amino acids do not fluoresce. If they are reacted with 9-fluorenylmethyl chloroformate, the products fluoresce and can be detected by fluorescence on an HPLC.

Sample Prep Schemes

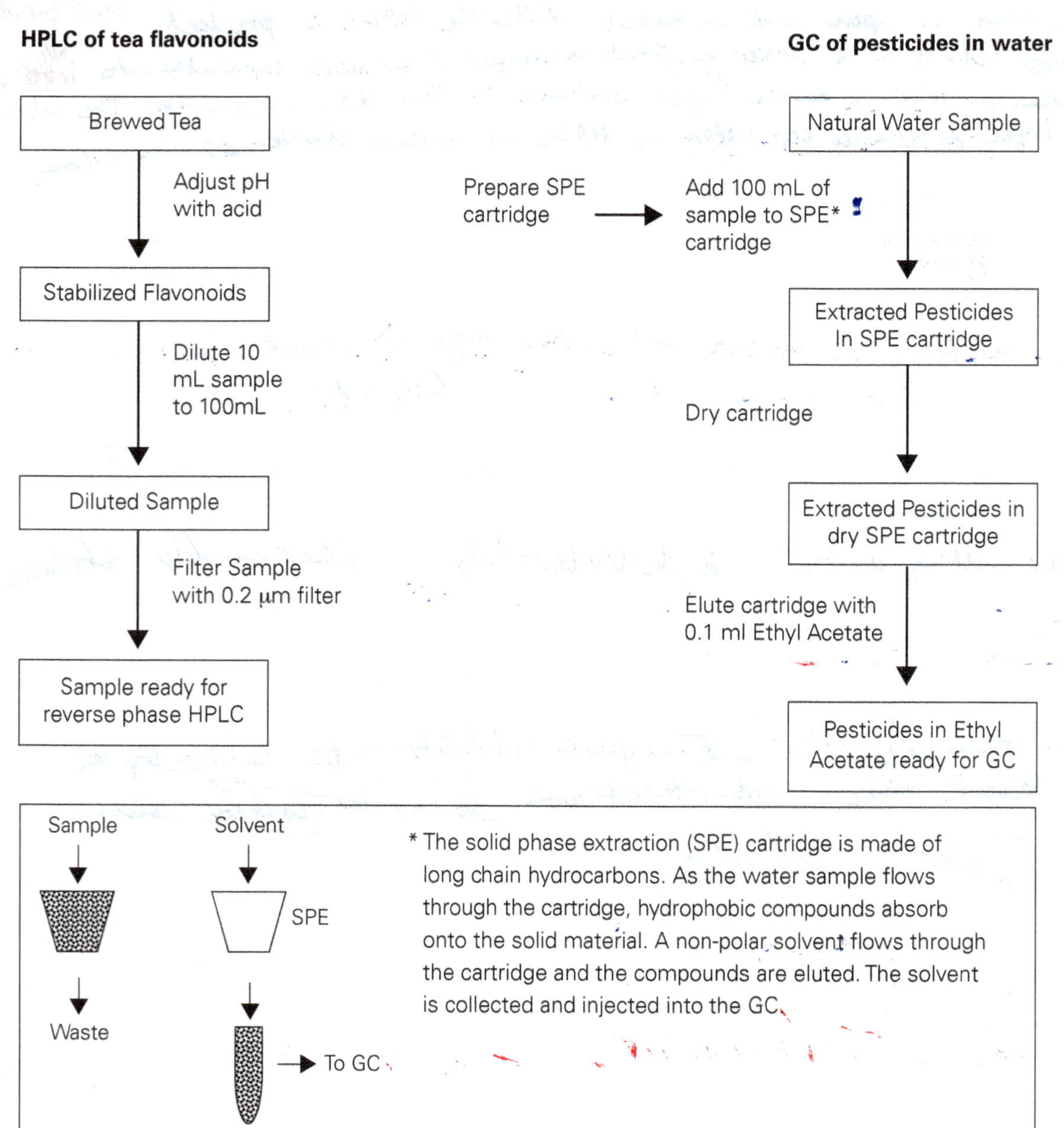

* The solid phase extraction (SPE) cartridge is made of long chain hydrocarbons. As the water sample flows through the cartridge, hydrophobic compounds absorb onto the solid material. A non-polar solvent flows through the cartridge and the compounds are eluted. The solvent is collected and injected into the GC.

Key Questions

6. Examine the sample preparation scheme for the HPLC of flavonoids. This scheme addresses two of the four purposes of sample prep. Describe which purposes are addressed and how it addresses these purposes.

2) adjust concentration

3) remove contaminants

7. Examine the sample prep scheme for the pesticides in water. Describe the change in concentration of the pesticides from the original water sample to the GC ready sample.

diluted by 100 mL

1000 more concentrated

8. Examine the HPLC sample prep scheme and determine the change in concentration for this particular type of sample.

$\frac{\text{undiluted concentration}}{10}$

9. How would you determine whether a sample requires dilution or concentration?

Run sample and compare to linear calibration curve of standards. See if it fits on the line.

depends on concentration range of instrument/detector

10. What would happen to polar contaminants in the water sample during the SPE extraction process?

They are trapped in the SPE cartridge with nonpolar walls. → waste

11. Both of these schemes involve samples in aqueous solutions. What purpose is necessary in the GC sample prep that is not necessary for reverse-phase HPLC sample prep?

1) change solvent 2) pH adjustment adjust concentration by concentration
3) Remove contaminants → SPE

12. Discuss with your group what three purposes are addressed in the GC sample prep scheme.

1) put analytes into right solvent
2) adjust concentrations
3) remove contaminants

13. The fourth purpose was not really addressed in either of these sample prep schemes. What properties of the analyte would make it difficult to analyze by GC (refer to Q5) and therefore require derivitization?

low volatility, high ~~melting~~ boiling point, low thermal stability

14. It is obvious the type of sample preparation necessary depends on the analytes and well as the sample matrix. Discuss with your group a set of guidelines that would determine the sample preparation necessary for both GC and reverse-phase HPLC. Record your group's consensus.

- Know / learn sample solubility in solvents of varying polarity
- anything w/ boiling point > 500°C → HPLC
- polar soluble → HPLC
- nonpolar soluble → GC

contaminants present?
dilution necessary?

Consider this...

Polynuclear aromatic hydrocarbons (PAH) are formed during combustion of fossil fuels, refuse burning, coke ovens, and even smoked foods. Several PAHs are carcinogens or potential carcinogens and therefore monitored in the environment. Shown in Figure 2 are chromatograms of a standard containing the 16 EPA controlled PAHs. Figure 2A is a capillary column GC chromatogram while Figure 2B shows an HPLC chromatogram. For both analyses the instrumental parameters are shown on the chromatogram.[1]

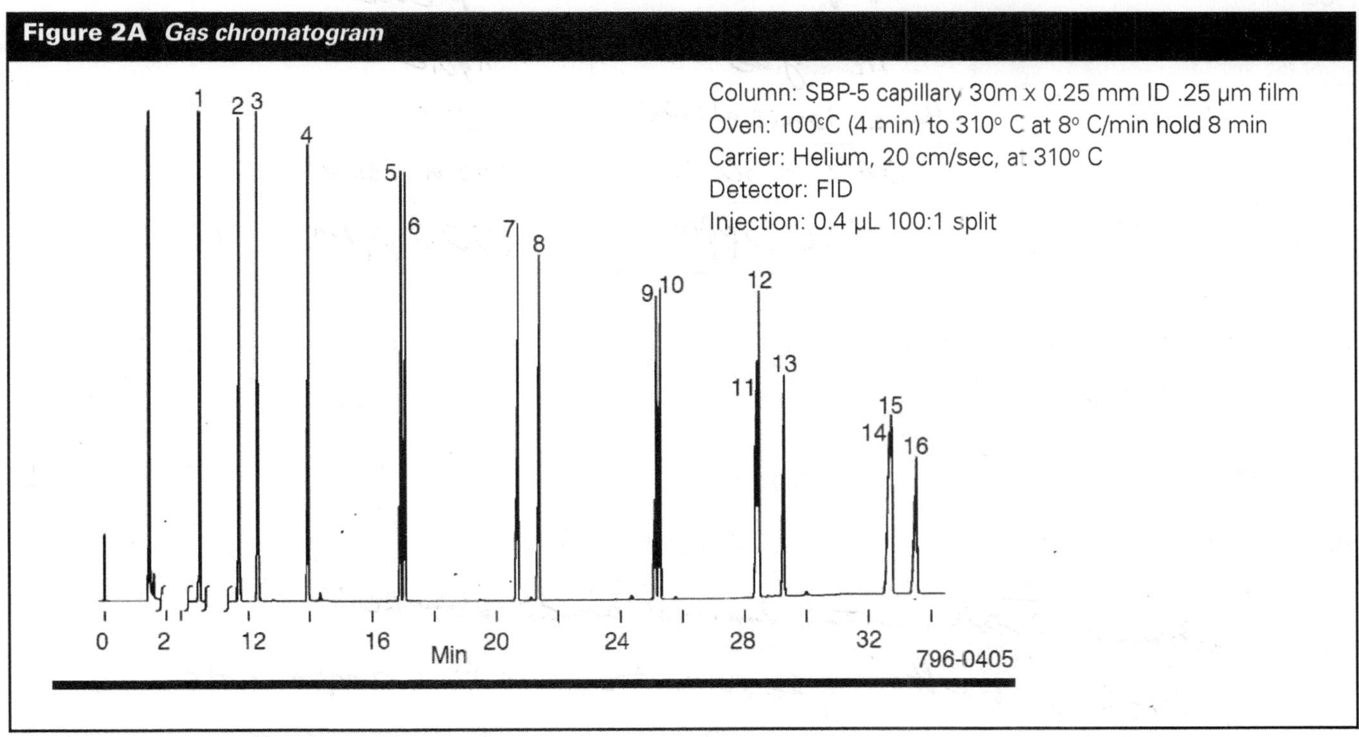

Figure 2A *Gas chromatogram*

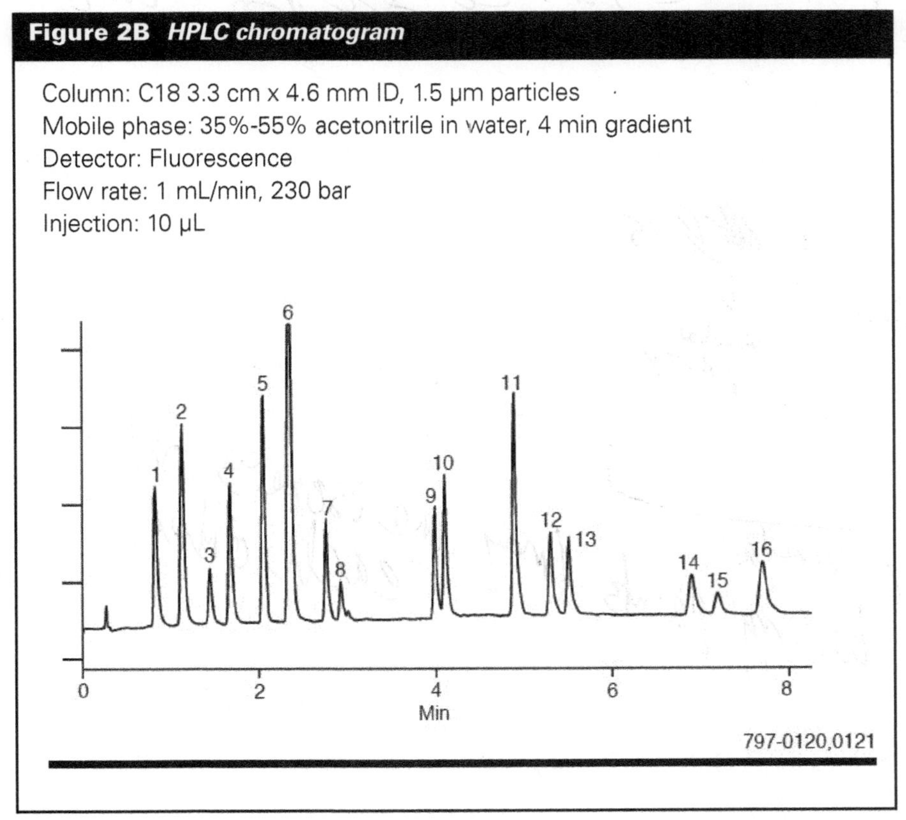

Figure 2B *HPLC chromatogram*

Concentration of PAHs (μg/mL)

	GC	HPLC
1. Naphthalene	1000	500
2. Acenaphthylene	1000	500
3. Acenaphthene	1000	1000
4. Fluorine	1000	100
5. Phenanthrene	1000	40
6. Anthracene	1000	20
7. Fluoranthene	1000	50
8. Pyrene	1000	100
9. Benzo(a)anthracene	1000	50
10. Chrysene	1000	50
11. Benzo(b)fluoranthene	1000	20
12. Benzo(k)fluoroanthene	1000	20
13. Benzo(a)pyrene	1000	50
14. Dibenzo(a,h)anthracene	1000	200
15. Benso(g,h,i)perylene	1000	80
16. Indeno(1,2,3-c,d)pyrene	1000	50

[1] Supelco, Application notes 108, 138, 1997.

Key Questions

15. Examine the instrument parameters used to produce each chromatogram. Fill in the table below for some of these parameters.

Instrument Parameters	GC	HPLC
Column	unpacked	packed
Mobile Phase	He gas	liquid
Detector	flame ionization	UV
Temp Ramp or gradient	60°C – 300°C	room temp
Concentration of acenaphthene	1000 ppm	1000 ppm

The following questions will examine three important criteria for chromatographic analysis: resolution, retention time, and sensitivity/detection limits.

Resolution

16. Which of the instruments provides better resolution of peaks for the PAHs analyzed? Discuss your choice with your group and defend your choice using specific examples.

~~GC, peaks are slimmer and sharper~~

HPLC, better separation of compounds 11 & 12 overlap in GC, as do 5 & 6 and 14 & 15

17. What pairs of peaks are particularly difficult to separate? Examine the table of PAH compounds at the end of this activity and describe why these pairs are problematic.

5&6, 11 & 12, 14 & 15

5&6 → similarity in polarity; 11 & 12 → similar structure, polarity; 14 & 15 → similar polarity

probably similar boiling points (many are isomers of each other)

The C-18 column used in this HPLC analysis is a long chain hydrocarbon stationary phase. The SPB-5 is a non-polar a general purpose column that provides an elution order based on boiling point.

18. Consider how the compounds interact with the stationary phase in both GC and HPLC. How does this explain why HPLC gives better separation for the PAH compounds?

HPLC separates by sample interaction with stationary phase, while GC separates by this plus boiling point. These don't necessarily complement each other and can be problematic together.

Retention time

19. Compare the retention times of the last compound to elute from the GC and HPLC. Why is this time important?

Significant difference in elution time.

longer run on GC

20. What are the advantages to a chemist of a short total analysis time?

You get to leave lab earlier.
More efficient use of time.

more income

Sensitivity/Detection Limit

While information concerning sensitivity and detection limits is not given for these analyses, we can get a sense of these parameters by comparing the mass of the compounds that were analyzed and the peaks on the chromatograms.

21. Using the concentration and the volume injected determine the mass of acenaphthene injected for each instrument.

Both are 1000 ppm

GC: $\frac{1000\,\mu g}{mL} \times 0.4\,\mu L$ = 400000 μg   0.4 μg

HPLC: $\frac{1000\,\mu g}{mL} \times 10\,\mu L$ = 10,000,000 μg   10 μg

check conversions

22. In the GC analysis, a 100:1 split is used in the injector. This means that a portion of the vapor in the injector is swept away and only 1/100 of this vapor travels into the column. This is essentially a 100 times dilution. Given this information what is the mass of the acenaphthene in the column?

$\frac{400{,}000\,\mu g}{100}$ = 4000 μg   0.004 μg   4 ng

23. Now compare the peak heights and areas of the acenaphthene in the GC and HPLC chromatograms. For these particular analyses, which would you infer has the lower detection limits?

~~HPLC~~ GC has lower detection limit → larger peak for 4 ug acenaphthene

24. Considering the three criteria examined here (resolution, retention time, and detection limit), discuss with your group which of these two methods you would use to analyze PAHs in a wastewater sample. Record your group's choice and reasoning.

GC has better detection limits and shorter ret. times

HPLC if worried about problematic compounds

Use HPLC; HPLC may have lower detection limit, but it also has much higher resolution and elutes compounds faster than GC with still better separation and elution.

25. What other criteria could be considered when choosing an instrumental method?

- price and cost of solvents/packing material, how much analyte is required
- Boiling points, solvent waste, complication and time for sample prep

26. Describe two examples of how your understanding of the differences between GC and HPLC has improved.

1) HPLC seems like it's better in almost every way than GC, given that analytes are more polar.

2) ~~[illegible]~~ GC uses strictly He gas or another carrier gas, while HPLC still uses solvents of varying polarity.

27. How are you more adept at interpreting schematic diagrams?

Not much more.

Applications

28. Highlight the differences between HPLC and GC instruments.

29. Since both HPLC and GC separate mixtures of compounds, why would a laboratory (i.e. a forensics lab) need both instruments?

30. Develop a checklist or set of questions about a sample to help determine if it would be best run on a GC or an HPLC.

31. Based on your understanding of instrument components (columns, detectors, injectors, etc) what could be changed in both the GC and HPLC to improve the analysis (resolution, retention times, or sensitivity) of PAHs?

32. Examine the following table of compounds and determine which might be suitable for GC analysis and which for HPLC (Some could be both).

Compound	Molar Mass (g/mol)	Boiling Point (°C)	Stability	Solubility	Instrument?
n-Butanol	130.23	195	Stable	Alcohol, ether, acetone	
Chlorobenzene	112.56	131	Stable	Alcohol, ether, benzene	
Pyrene	202.26	393	Stable	Benzene	
Citric Acid	192.12	---	Decomposes	Water, alcohol	
Vitamin D	384.64	---	Decomposes	Alcohol, ether	

33. The flame ionization detector (FID) in GC and the UV detector in HPLC both respond to many organic compounds and therefore can be used for a variety of mixtures. They do not, however, give any structural information about the compounds. In some cases different compounds can have the same retention time, therefore retention time alone is not an absolute identifier of a chemical compound. Using additional resources, find 3 different detectors for GC and/or HPLC that can be used to identify the compounds after separation and discuss the advantages of each of these detectors.

Structures of Polyaromatic Hydrocarbons[2]

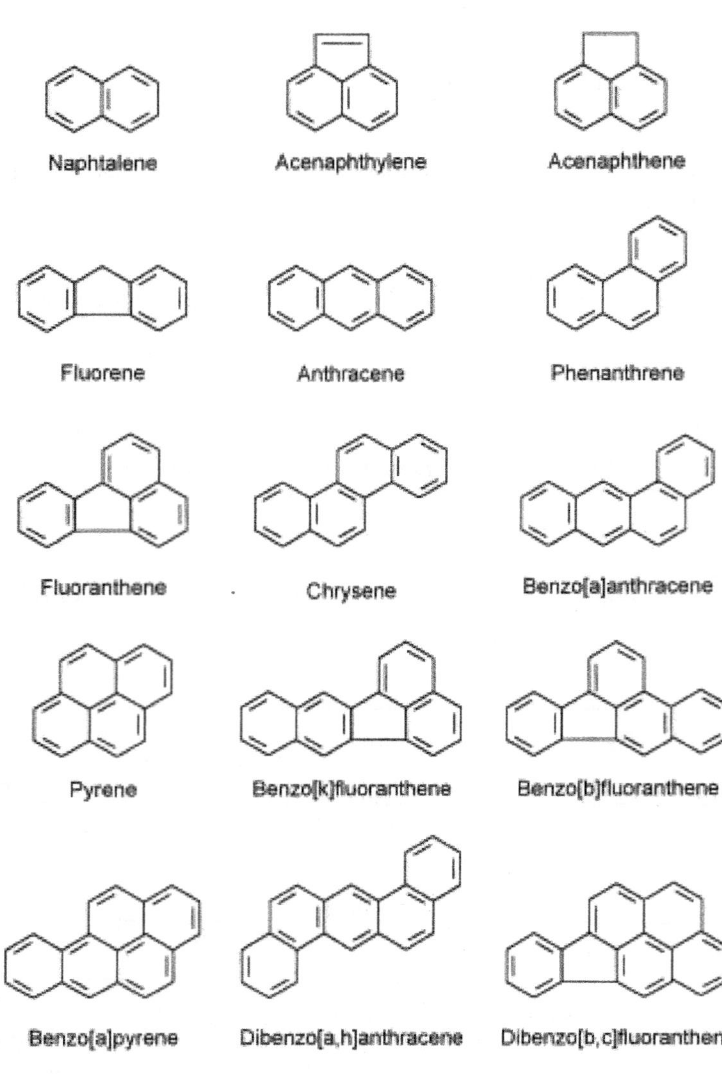

[2] Journal of Chromatography A, vol. 885, issues1-2, 14 July 2000, p.273-290.

Introduction to Gel Filtration Chromatography

Learning Objectives

Students should be able to:

Content

- Explain how the sample protein characteristics and the characteristics of both the stationary phase and the mobile phase impact elution.
- Use their understanding of how proteins interact with gel filtration resins to derive an expression for the term $k_{av.}$

Process

- Relate molecular interactions and macroscale processes. (Critical thinking)
- Derive mathematical expressions that describe the separation process. (Critical thinking)

Prior knowledge

- An understanding of intermolecular forces at the general chemistry level.

Further Reading

- Harris, D. C. 2007. *Quantitative Chemical Analysis,* 7th Edition, Section 23-2, p.506. New York: W.H. Freeman.
- Skoog, D. A., F. J. Holler, and S. R. Crouch. 2007. *Principles of Instrumental Analysis,* 6thEdition, 26A-B, pp.763-768. Belmont, CA: Thomson Brooks/Cole,.
- *Gel Filtration: Principles and Methods,* Edition A1, Handbook #18-1022-18. Piscataway, NJ: Amersham Biosciences.
- Voet, D, J. G. Voet, and C. W. Pratt, 2008. *Fundamentals of Biochemistry,* 3rd Edition. Chapter 5. Hoboken, NJ: John Wiley & Sons.

Author

Caryl Fish, Mary Walczak, Ruth Riter, Paul Jackson, and Kathleen Cornely

Consider this…

A chromatographic separation is shown in Figure 1 in which a glass column is packed with a solid gel filtration medium suspended in buffer. The solid material, sometimes referred to as a resin, is in a bead form. The solid beads and the buffer enclosed by the beads constitute the stationary phase. The buffer bathing the beads is the mobile phase. A sample containing two proteins labeled A and B is loaded on top of the column and buffer is allowed to flow through the column. The two proteins are partitioned between the stationary and mobile phases.

Figure 1 ***Separation of proteins A (large gray spheres) and B (small gray spheres) by gel filtration chromatography***

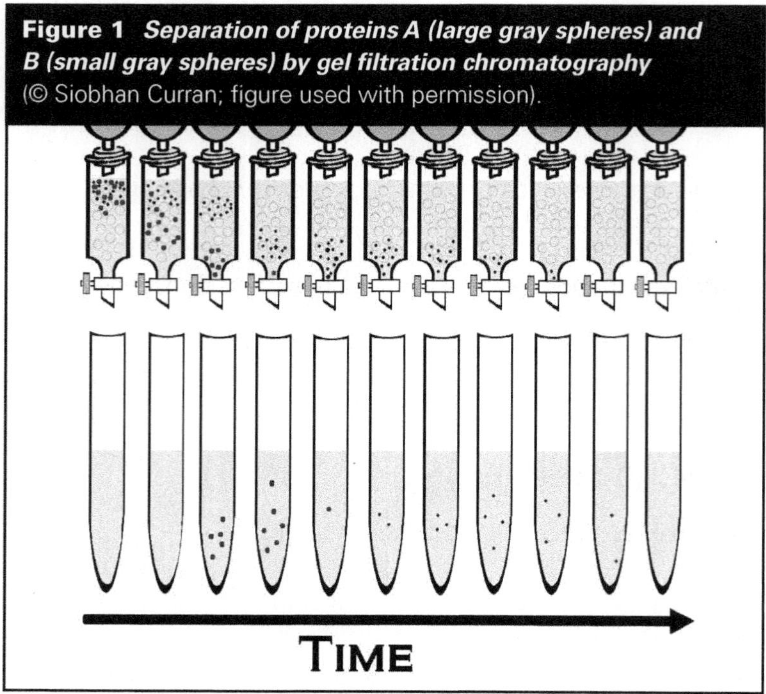

Key Questions

1. Which protein elutes from the column first, A or B?

The elution volume (V_e) for a solute describes the volume of buffer required to elute a component of the sample.

2. Which protein requires a larger elution volume, A or B?

The effluent buffer emerging from the column passes through a UV monitor set at 280 nm. Proteins absorb ultraviolet light at this wavelength; the absorbances are shown as peaks on the elution profile shown in Figure 2. Fractions are collected; the fraction collector records a tic mark when a fraction is filled. Each fraction contains a volume of 10 mL.

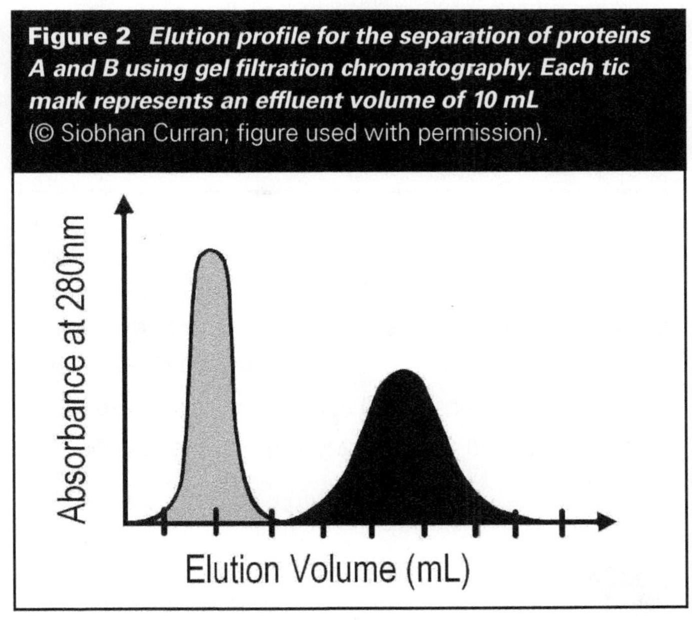

Figure 2 ***Elution profile for the separation of proteins A and B using gel filtration chromatography. Each tic mark represents an effluent volume of 10 mL*** (© Siobhan Curran; figure used with permission).

3. Which peak on the elution profile corresponds to protein A and which peak corresponds to protein B? Label each peak on the chromatogram as either A or B. Does your group agree with these labels?

4. The elution profile is used to determine V_e for each protein eluting from the column. The V_e determination should be consistent from person to person. What part of the peak should be used to give the most consistent V_e? Explain your reasoning to your group.

5. Use the elution profile in Figure 2 to determine the V_e for each protein eluting from the column and display your answers in the table below.

Protein	Elution volume, V_e
A	
B	

Consider this...

The gel filtration resin consists of a gel matrix of inert porous beads. The two proteins are partitioned between the stationary phase and the mobile phase based on the abilities of the proteins to diffuse in and out of the pores of the beads. A diagram of a single bead is shown in Figure 3. Proteins may have full access, partial access, or no access to the interior of the gel matrix, depending on whether they are able to penetrate the pores of the bead to enter the gel matrix.

Figure 3 ***A diagram of a single bead in a gel filtration matrix.***

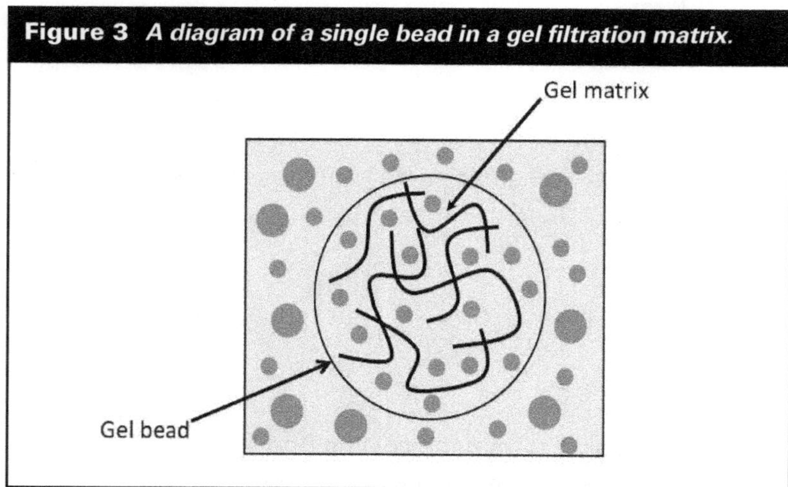

6. Examine the distribution of proteins between the mobile phase (outside of the bead) and the stationary phase (inside of the bead). Which protein has a higher percentage of molecules in the mobile phase?

7. Refer to your answers to Q1 and Q6 to explain why the two proteins elute from the column in the order observed. Come to a group consensus before proceeding.

8. On what basis are components separated in gel filtration chromatography?

Consider this...

The important terms in gel filtration chromatography are defined below:

- V_t is the total column volume and is easily calculated from the height and radius of the column using the formula of a cylinder, $V_t = \pi r^2 h$.

- V_o is the volume of the mobile phase and is also called the void volume.

- V_s is the volume of the stationary phase and denotes the volume of the buffer inside the beads.

Note that this is only an estimate because the volume of the gel particles themselves is also a part of the stationary phase; however this is ignored both because the value is difficult to determine and because this portion of the stationary phase is inaccessible to all solutes.

Figure 4 ***Schematic diagrams illustrating various volume concepts in gel filtration chromatography*** (© Siobhan Curran; figure used with permission).

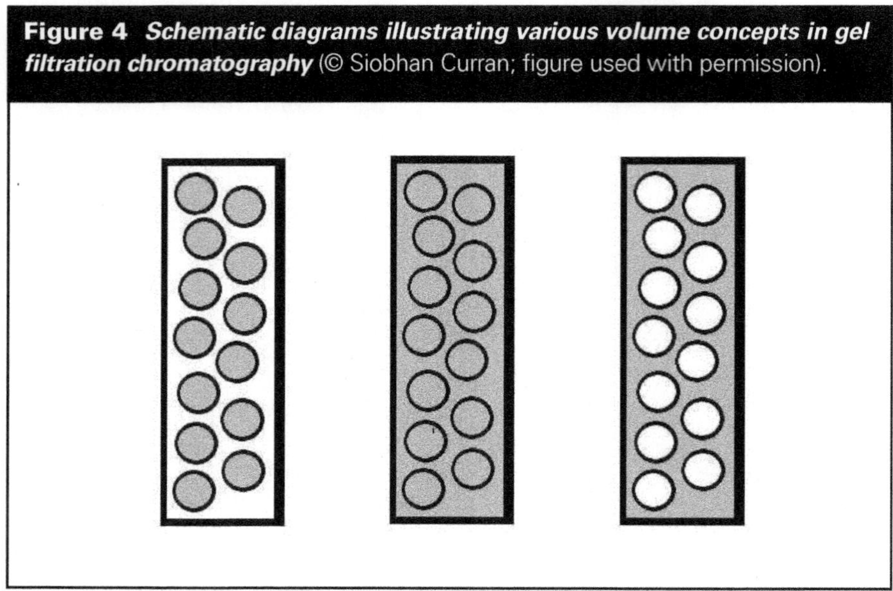

9. Schematic diagrams depicting the volume terms described above are shown in Figure 4. Use the descriptions to match each diagram with its proper term. Note that the volumes described are indicated by the shaded regions in Figure 4.

10. Determine the total volume, V_t, for a column with a length of 20.0 cm and a diameter of 2.5 cm.

11. Refer to Figure 4. Which volume is the only volume that the largest proteins have access to, and why?

12. Refer to Figure 4 and to the elution profile shown in Figure 2. How could you experimentally determine a value for the volume described in Q11? What is the value for this volume for the gel filtration column illustrated in Figure 1?

13. Summarize the results of your calculations in the table below.

Elution volume, V_e, for Protein A	
Elution volume, V_e, for Protein B	
Total column volume, V_t	
Void volume, V_o	

14. Using Figure 4 and what you've learned, write an equation to calculate an estimate of V_s.

The term V_{sp} is used to describe the volume of the stationary phase in which a particular solute spends its time.

15. Consider V_{sp}, as described above. Draw a picture, similar to the column diagrams shown in Figure 4, which illustrates the concept of V_{sp} for Protein B.

16. If the elution volume, V_e, of a solute is the volume that the solute has access to in <u>both</u> the stationary phase <u>and</u> the mobile phase, how could you mathematically determine V_{sp}? Use both the information provided in Figure 4, and the picture you drew in Q15, to write an equation that you could use to calculate V_{sp}.

The term k_{av} represents the fraction of the total stationary phase that is available for a given solute in a gel filtration column.

17. Locate the equation you wrote earlier in this activity that describes the volume of the portion of the stationary phase accessed by the solute. Rewrite the equation below.

Locate the equation you wrote earlier in this activity that describes the total volume of the stationary phase. Rewrite the equation below.

Use the definition of k_{av} provided, and the equations that you wrote above, to write an equation that you can use to calculate the value of k_{av} for each solute separated by a gel filtration column. Write your equation individually and then compare answers amongst the members of the group.

18. Using the equation you wrote in Q17, and the values tabulated in Q13, calculate the values of k_{av} for proteins A and B.

How do the values of k_{av} for proteins A and B relate to their elution volumes as shown in Figure 2? As individuals, write a statement that summarizes this relationship, then compare statements amongst group members.

The retention coefficient R is the ratio of V_e to V_0 for a given solute.

19. Calculate the retention coefficients for Proteins A and B.

How do the values of R for proteins A and B relate to their elution volumes? As individuals, write a statement that summarizes this relationship, then compare statements amongst group members.

20. Summarize the results of your calculations by completing the table below.

	Protein A	**Protein B**
Elution volume, V_e		
k_{av}		
R		

21. Are there maximum and minimum allowable values for k_{av}? If so, what are they?

22. Are there maximum and minimum allowable values for R? If so, what are they?

23. How confident are you in predicting the elution order of compounds from a gel filtration column? In calculating and interpreting values for k_{av}? If you are not confident, what additional questions do you have?

24. What is one way this activity improved your ability to visualize the microscopic separation processes imbedded in the appearance of an elution profile?

Application

25. Gel filtration chromatography is often used as a desalting technique during a protein purification process. Why is this technique particularly effective at removing salt from a protein or a mixture of proteins?

26. A *selectivity curve* can be plotted by graphing k_{av} versus log molecular weight for a set of standard proteins. The plot is usually linear for k_{av} values in the range of 0.1 to 0.7. The linear portion of the selectivity curve is used to determine the *fractionation range* and the *exclusion limit* for a bead with a specific pore size. The fractionation range is the molecular weight range of the proteins that can be successfully separated from one another using that particular gel filtration resin. These proteins have some access to the stationary phase. The *exclusion limit* is the minimum size of the protein that has no access to the stationary phase; these proteins are *excluded* from entering the beads. A selectivity curve is shown in Figure 5.

Figure 5 *A selectivity curve can be used to determine the fractionation range for a particular gel filtration resin.*

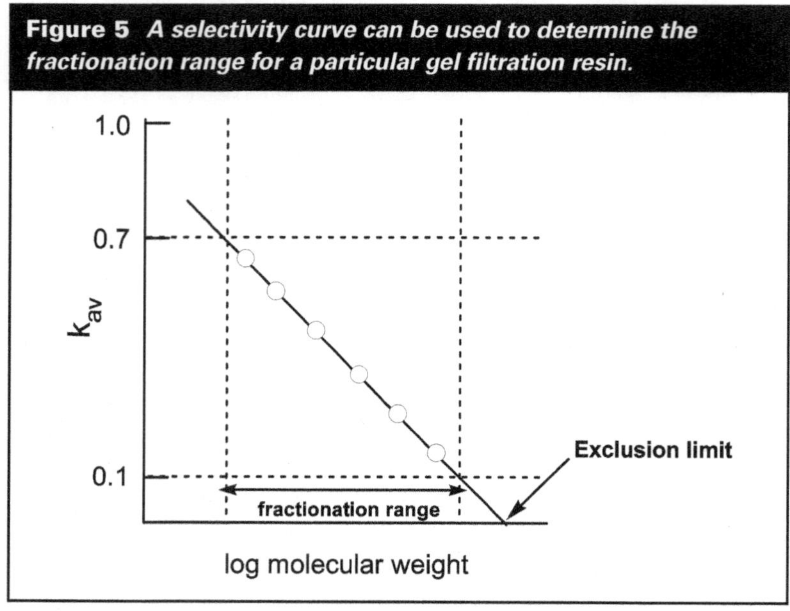

Figure 6 *The selectivity curve for Superdex-75.*

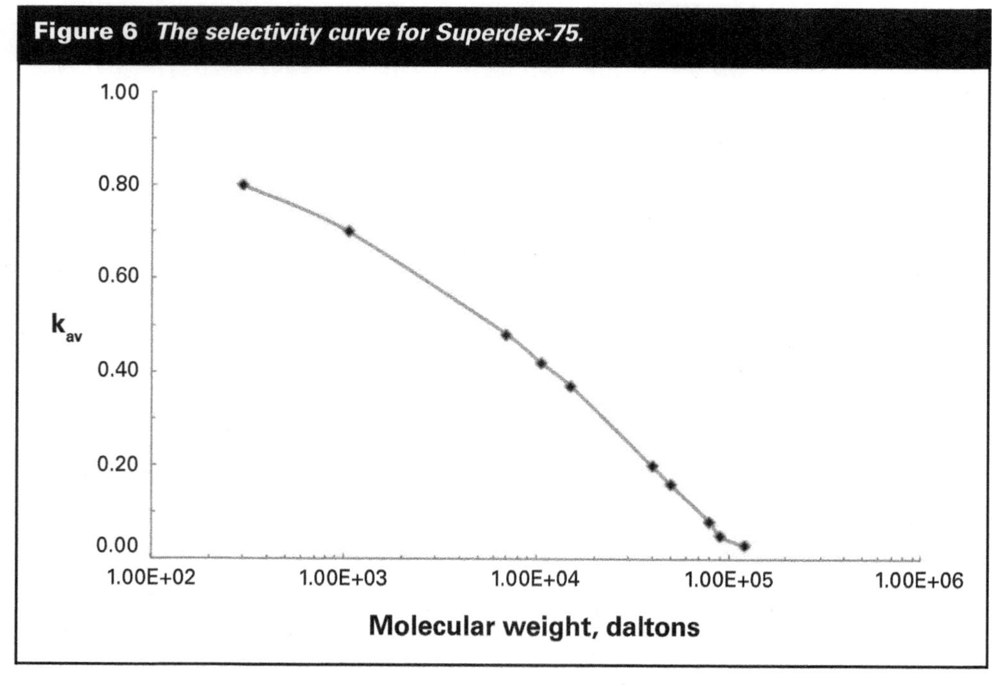

A selectivity curve for the gel filtration resin Superdex-75 is shown in Figure 6. What is the fractionation range for Superdex-75? What is its exclusion limit?

27. An SDS-PAGE gel showing the separation of the major proteins in the red blood cell membrane is shown in Figure 7. Predict the results that would be obtained if a mixture of these proteins was loaded onto a Superdex-75 column. Would this resin separate the proteins effectively? Explain.

Figure 7 ***Separation of the major proteins in red blood cell membranes by SDS-PAGE. Note that the migration of glycophorin is slowed due to its extensive glycosylation.***

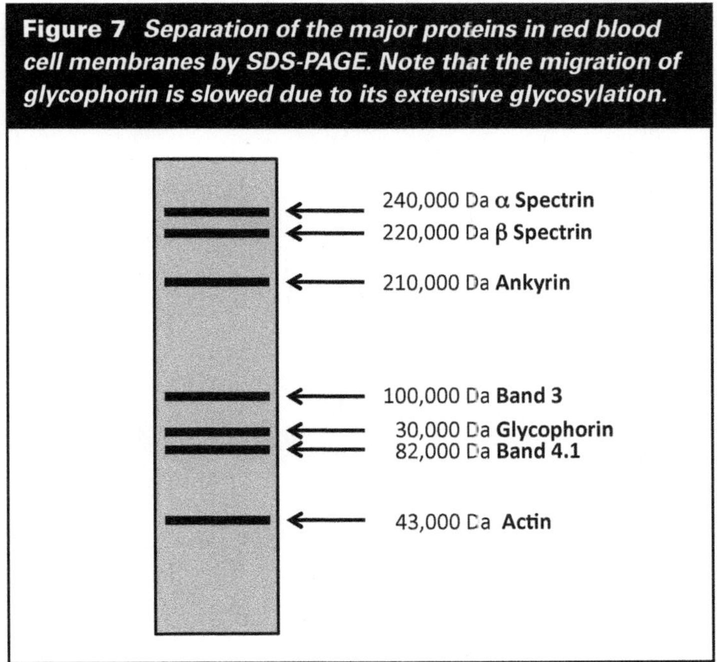

28. Proteins from RNA tumor viruses that differ in molecular weight by as little as 15% were successively separated by gel filtration chromatography when guanidinium hydrochloride was included in the stationary phase buffer. Guanidinium hydrochloride is a denaturing agent that unfolds proteins. Why was the inclusion of the denaturing agent in the buffer critical to the successful separation of these proteins? (Green, R. W., and D. P. Bolognesi. 1974. "Isolation of Proteins by Gel Filtration in 6 M Guanidinium Chloride: Application to RNA Tumor Viruses." Analytical Biochemistry 57: 108-117).

ANA-POGIL Project Activities: Topics and Learning Goals

Analytical Tools		Quant	Instrumental	Environmental	Bio-analytical
Content Goals	**Process Skills**				
Accuracy, Precision and Tolerance: Sorting Out Glassware		✓		✓	✓
• Differentiate between accuracy and precision. • Compare the tolerance of various pieces of glassware. • Classify data in terms of accuracy and precision	• Interpret tables of volumetric glassware tolerances (Information processing) • Infer based on tabulated data (Critical thinking) • Collaborate with group members (Teamwork)				
Solutions and Dilutions		✓		✓	✓
• Design a procedure for making a particular solution and assess the advantages of different approaches • Choose the appropriate glassware to ensure the desired level of precision of a particular solution • Convert between different concentration units (e.g., ppm to M)	• Develop alternative pathways for diluting solutions (Information processing) • Design approaches for preparing solutions (Problem solving) • Infer chemical processes based on reactions (Critical thinking)				
Classical Analytical Methods – A Design Perspective On Volumetric Measurement		✓			
• Describe the characteristics of the chemical substances and reactions used in volumetric analysis, particularly as they affect accuracy and precision. • Explain the rationale for the procedural steps in volumetric analysis, particularly as they affect accuracy and precision. • Explain how visual indicators are used • Explain the role of primary standard materials.	• Apply chemical concepts and knowledge of mass and volume measurement to making design decisions about volumetric measurement (Critical thinking) • Use typical titration measurement calculations (Problem solving)				
Sample Preparation		✓	✓	✓	
• Evaluate sample preparation methods for completeness. • Determine the concentration of analyte in the sample from the concentration in the digestate. • Explain how to assure sample integrity with respect to contamination and accuracy.	• Form an understanding of the accuracy of a measurement (Critical thinking) • Analyze methods for robustness and sources of error, and overall quality (critical thinking)				
Instrumental Calibration		✓	✓	✓	✓
• Use a correlation coefficient to ascertain the fit of data to a calibration curve • Determine the minimum detectable concentration and lower limit of quantitation and distinguish between these. • Determine the linear dynamic range from a set of calibration data.	• Interpret tabulated data (Information processing) • Produce an xy graph of data (Graphing) • Interpret calibration curves (Information Processing)				

Analytical Tools		Quant	Instrumental	Environmental	Bio-analytical
Quality Assurance Measures – How Good is the Data?		✓	✓	✓	
• Calculate the spike recovery and relative deviation between replicate samples. • Use spike recovery and relative deviation between replicate samples to assess and alidate sample analysis quality. • Use spike recovery and relative deviation between replicate samples to assure accuracy and precision of an analysis.	• Validate data (Problem solving) • Interpret tabulated data (Information processing)				
Instrumental Calibration: Method of Standard Addition		✓	✓		
• Recognize when the method of standard addition is needed (i.e. matrix effects) • Distinguish between the method of standard additions and normal calibration using standards. • Determine the concentration of an unknown in an original sample using the method of standard additions.	• Interpret calibration graphs (Information processing) • Calculate concentrations of unknown solutions from standard addition data (Information processing) • Read and interpret a standard method (Information processing)				
Instrumental Calibration: Method of Internal Standards		✓	✓	✓	
• Describe calibration by internal standards as a method to deal with irreproducible instrumental results • Calculate the concentration of an analyte in a sample when analyzed using the method of internal standards. • Determine the need for the internal standard method.	• Construct and interpret graphs. (Information processing)				
Interlaboratory Comparisons		✓	✓	✓	
• Explain how random and systematic error affects experimental results. • Describe what the purpose of an interlaboratory study is and how it might be conducted. • Explain how spike recovery may be used to evaluate accuracy of a method and of a laboratory's performance.	• Draw inferences from the data presented in graphical form (Information processing)				
Sampling Error and Sampling Designs		✓	✓	✓	
• Describe a sample as a part taken from the whole system of interest, where the part is representative of the chemical composition of the whole. • Describe the different types of sampling designs and how they address compositional or temporal heterogeneity in the system of interest. • Estimate the relative importance of sampling error relative to measurement error.	• Seek patterns in tabulated data and summarize trends (Information processing) • Design a sampling strategy for a practical case (Problem solving)				

Statistics		Quant	Instrumental	Environmental	Bio-analytical
Content Goals	**Process Skills**				
Errors in Measurements and Their Effect on Data Sets		✓			
• Classify different sources of experimental error and determine their effect on the mean and dispersion of the data. • Develop the ability to judge whether an analytical balance has been used correctly. • Define situations in which it is reasonable to reject data.	• Develop critical skills in assessing the quality of data (Critical thinking) • Identify qualitative difference among data sets (Information processing) • Develop generalizations from tabulated data (Information Processing)				
The Gaussian Distribution		✓	✓		
• Explain how the shape and position of a Gaussian Distribution are controlled by the mean and standard deviation. • Explain how histograms can be used to illustrate the frequency and distribution of data, and how the sample size affects the accuracy of the histogram. • Describe the relationship between the Gaussian distribution function and probability.	• Interpret graphs of data, especially histograms. (Information processing) • Estimate the distribution parameters of a data set (Information processing) • Draw conclusions about data sets. (Critical thinking)				
Statistical Tests of data: The t Test		✓		✓	
• Explain how the t value changes due to changes in the means, the standard deviations, and the number of data. • Explain how the t test is used to make decisions regarding significance, and be able to use Excel's t test output to make such decisions. • Develop an understanding of the t value's relationship to the probability of a significant difference.	• Interpret tables of data. (Information processing) • Draw conclusions about data sets. (Critical thinking) • Express concepts in grammatically correct sentences. (Communication)				
Statistical Tests of Data: The F Test		✓			
• Explain how the F value changes with respect to the standard deviations of the data sets. • Explain how the F test is used to make decisions regarding significance, and be able to use Excel's F test output to make such decisions. • Explain how the results of the F test will affect the t test.	• Interpret tables of data. (Information processing) • Draw conclusions about data sets. (Critical thinking) • Express concepts in grammatically correct sentences. (Communication)				
Linear Regression for Calibration of Instruments		✓	✓	✓	
• Describe linear regression as an averaging process • Interpret the results of a linear regression analysis • Evaluate the quality of a linear model for calibration of an instrumental measurement	• Construct and interpret graphs (Information processing)				

Equilibrium		Quant	Instrumental	Environmental	Bio-analytical
Content Goals	**Process Skills**				
The Importance of Ionic Strength		✓			
• Be able to describe the role of electrolytes in determining the ionic strength of solutions. • Be able to explain, on a microscopic level, the effect of ionic strength on equilibrium concentrations. • Be able to calculate the ionic strength of a solution and predict its impact on equilibrium concentrations.	• Identifying key properties of electrolyte solutions (Problem solving) • Depicting a shared understanding of microscopic solution interactions (Information processing, teamwork) • Challenging assumptions (Critical thinking)				
The Importance of Ionic Strength – Biochemistry Focus					✓
• Use the concept of ionic atmosphere to explain how ions behave in solution. • Be able to calculate the ionic strength of a buffer solution. • Compare the relationship between ionic strength and concentration for various buffers.	• Identify key properties of electrolyte solutions (Problem solving) • Form a shared understanding of microscopic solution interactions (Teamwork) • Challenging assumptions (Critical thinking)				
Activity and Activity Coefficients		✓			
• Describe the impact of ionic strength on activity coefficients and the equilibrium concentrations of slightly soluble salts. • Develop strategies for calculating activity coefficients, activities and equilibrium concentrations in various electrolyte solutions. • Discern the ionic strength range for which activities need to be used in equilibrium calculations.	• Finding, predicting trends in data (Information Processing) • Identifying similarities and differences in solutions (Critical thinking) • Estimating values, choosing solving strategies for various concentration regions (Problem solving)				
Multiple Equilibria: When Reactions Compete		✓		✓	
• Explain how a competing equilibrium affects the solubility of a sparingly soluble compound in water. • Predict the pH dependence of the solubility of a sparingly soluble salt.	• Identifying key reactions (Problem solving) • Challenging assumptions (Critical thinking) • Predicting pH dependence (Information processing)				
pH of Solutions of Strong Acids and Bases		✓		✓	
• Calculate the pH of solutions containing any concentration of strong acid or base.	• Recognize when the result of a calculation makes sense (Critical thinking) • Identify assumptions (Problem solving)				
Acid-base Distribution Plots		✓		✓	
• Describe how the molar concentrations of mono- and polyprotic weak acids and their conjugate bases vary with pH. • Identify the principal species resulting from the dissociation of a weak acid at a given pH.	• Sketch ionic distribution graphs given acid-base parameters (Critical thinking, visualization) • Interpreting graphs (Information processing)				

Equilibrium

Content Goals	Process Skills	Quant	Instrumental	Environmental	Bio-analytical
Acid-Base Distribution Plots – Biochemistry Focus					✓
• Describe how the molar concentrations of mono- and polyprotic weak acids and their conjugate bases vary with pH. • Use an acid-base distribution plot to determine the pKa values of weak acids and to determine the appropriate buffering region for a weak acid/ conjugate base system. • Use the pKa value of the functional groups of amino acids to estimate a value for the pI of an amino acid.	• Sketch ionic distribution graphs given acid-base parameters (critical thinking– visualization) • Interpret graphs (information processing)				
The Acid-Base Distribution Functions		✓			
• Explain what information acid-base fractional distribution functions provide. • Use acid-base distribution functions to calculate the concentration of a species arising from any weak acid at a given pH.	• Extend the application of a method to new situations using pattern recognition (Critical thinking) • Work effectively with others to reach a consensus (Teamwork) • Evaluate answers for reasonableness (Critical thinking)				
The Buffer Zone: What is a Buffer and in What pH Range is it Effective?		✓		✓	
• Explain how the concentration of buffer components determines the pH of the solution. • Explain the effects of additions of strong acids/ bases to buffered and non-buffered solutions • Determine the relationship between the pKa of an acid and the optimal pH range of a buffer in order to select an appropriate buffer for a particular pH range.	• Interpret tabulated information (Information processing) • Recognize and predict trends in data (Critical thinking) • Generalize problem solutions (Problem solving and critical thinking) • Include all group members (Teamwork)				
The Buffer Zone — Biochemistry Focus					✓
• Explain how the pH of a solution is altered by the concentration of buffer components. • Determine the relationship between the pKa of an acid and the optimal pH range of a buffer in order to select an appropriate buffer for a particular pH range. • Calculate the buffer capacity of a solution.	• Interpret tabulated information (Information processing) • Recognize and predict trends in data (Critical thinking) • Generalize problem solutions (Problem solving and critical thinking) • Include all group members (Teamwork)				
When Acids and Bases React: Laboratory		✓			✓
• For a given acid/base reaction, distinguish between equivalence point and endpoint. • Given a titration curve, identify whether a monoprotic acid is "strong" or "weak". • Given a titration curve, identify the main features of the curve, i.e. (1) whether an analyte is an acid or base, (2) whether an analyte is a mono- or polyprotic acid, and (3) what is/are the analyte's pK value(s).	• Obtain titration curve using pH probe, drop counter, and data acquisition software (Experimental technique) • Determine the equivalence point of a titration curve from 1st and 2nd derivative plots of the curve using computer software, i.e. Excel, Vernier software, etc. (Information processing) • Predict the appearance of a titration curve (critical thinking)				

Electrochemistry		Quant	Instrumental	Environmental	Bio-analytical
Content Goals	**Process Skills**				
Electrochemistry: The Microscopic View of Electrochemistry		✓			
• Describe microscopic-level processes occurring at electrodes and explain how these processes result in formation of the electrochemical double layer. • Explain the origins of electrochemical potential. • Use E° values to predict the favored reaction in electrochemical reactions.	• Interpret pictures of electrochemical processes. (Information Processing) • Draw and interpret microscopic pictures of electrochemical processes. (Information Processing) • Draw and interpret graphical representations of electrochemical processes. (Critical Thinking)				
Electrochemistry: Calculating Cell Potentials		✓			
• Apply the criteria for an electrochemical cell in the standard state to determine if a cell is initially in the standard state. • Predict how the ion concentrations in an electrochemical cell will change as an electrochemical cell circuit is completed and allowed to run. • Determine Ecell for a given electrochemical cell and decide whether the reaction is spontaneous in the forward or reverse direction.	• Interpret drawings of electrochemical cells (Information processing) • Compare and contrast different electrochemical cells. (Critical thinking)				
Spectrometry					
The Beer-Lambert Law		✓	✓	✓	
• Identify each term in the Beer-Lambert law and explain its effect on absorbance. • Explain the relationship between transmittance and absorbance. • Use Beer's Law in quantitative measurements.	• Prepare and interpret graphs (Information processing) • Develop mathematical expressions to describe data (Problem solving).				
The Beer-Lambert Law – Biochemistry Focus					✓
• Identify each term in the Beer-Lambert law and explain its effect on absorbance. • Explain the relationship between transmittance and absorbance. • Use Beer's Law in quantitative measurements.	• Prepare and interpret graphs (Information processing) • Develop mathematical expressions to describe data (Problem solving).				
Atomic and Molecular Absorption Processes		✓	✓		
• Elate the wavelength and energy of light absorbed to molecular and atomic transitions. • Compare molecular and atomic absorption processes and explain differences in bandwidth in the resulting spectra.	• Compare and contrast spectra and absorption processes (Critical thinking)				

Introduction to Spectrometers			✓		
• Identify the components of a UV-Vis spectrophotometer. • Describe characteristics of the source, wavelength selector, sample holder, and detector in a UV-Vis spectrophotometer. • Compare different types of instrument components based on these key characteristics and the goals of the instrumental analysis.	• Interpret graphical data. (Information processing) • Interpret schematic diagrams and relate to real instruments. (Information processing) • Critical thinking by comparison and contrast.				
Infrared Spectroscopy: Vibrational Modes			✓		
• Classify a vibration as a bend or a stretch. • Predict the number and type of motions for a particular molecule. • Explain the relationship between Infrared (IR) absorbance and changes in dipole moment.	• Use the chemical education digital library to view molecular vibrations and corresponding IR spectra • Infer general relationships from specific examples (Generalizing) • Specify the frequencies of some infrared vibrational modes (Integrating prior knowledge)				
Mass Spectrometry - Interpretation			✓		
• Explain the source of fragment ions in electron impact mass spectra. • Identify the base peak and parent (molecular ion) in a mass spectrum. • Recognize patterns in the mass spectra and use them to identify functional groups.	• Comparing and contrasting graphical data (critical thinking) • Analyze graphical information (Information processing)				

Chromatography and Separation

Intro to chromatography		✓	✓	✓	
• Relate the separation process to features of the chromatogram, such as retention time, and resolution. • Explain how the chemical interactions between a solute and both the stationary phase and the mobile phase impact retention. • Predict the elution order for a set of compounds given the mobile and stationary phase composition.	• Relate molecular interactions and macroscale processes (Critical thinking)				
Band Broadening Effects in Chromatography		✓	✓		
• Explain how band broadening (i.e band dispersion) is related to linear flow rate. • Discuss how the three components in the van Deemter equation affect bandwidth.	• Given a plot of data, identify the mathematical relationship (Critical thinking) • Sketch graphs of known mathematical relationships (Information processing)				
Gas Chromatography or HPLC, Which do You Choose?		✓	✓	✓	
• Describe the differences and similarities between Gas Chromatography (GC) and High Performance Liquid Chromatography (HPLC). • Compare sample preparation for GC and HPLC. • Compare resolution, sensitivity, and analysis time for a particular chromatographic separation.	• Interpreting schematic diagrams (Information Processing) • Identifying similarities and differences (Critical thinking) • Integrating prior knowledge to choose appropriate instrument (Critical Thinking)				

Introduction to Gel Filtration Chromatography		✓	✓		✓
• Explain how the sample protein characteristics and the characteristics of both the stationary phase and the mobile phase impact elution. • Use their understanding of how proteins interact with gel filtration resins to derive an expression for the term kav.	• Relate molecular interactions and macroscale processes (Critical thinking) • Derive mathematical expressions that describe the separation process. (Critical thinking)				

ANA-POGIL Activities & Analytical Texts Mapping

Activity	Harris Chapter(s) 8th ed.	Skoog Chapter 6th ed.	Christian Chapter 9th ed.
Analytical Tools			
Accuracy, Precision, and Tolerance: Sorting Out Glassware	2 (Tools of the Trade) 3 (Experimental Error)	Appendix 1	3 (Statistics and Data Handling in Analytical Chemistry)
Solutions and Dilutions	1 (Chemical Measurements) 2 (Tools of the Trade)	N/A	5 (Stoichiometric Calculations: The workhorse of the Analyst)
Classical Analytical Methods - A design Perspective on Volumetric Measurement	1 (Chemical Measurements)	N/A	8 (Acid-Base Titrations)
Sample Preparation	27 (Sample Preparation)	N/A	18 (Sample Preparation: Solvent and Solid-phase Extraction)
Instrumental Calibration	4 (Statistics) 5 (Quality Assurance and Calibration Methods)	1 (Introduction)	3 (Statistics and Data Handling in Analytical Chemistry)
Quality Assurance Measures – How Good is the Data?	5 (Quality Assurance and Calibration Methods)	1 (Introduction)	4 (Good Laboratory Practice: Quality Assurance and Methods of Validation)
Instrumental Calibration: Method of Standard Addition	5 (Quality Assurance and Calibration Methods)	1 (Introduction)	17 (Atomic Spectrometric Methods)
Internal Standards	5 (Quality Assurance and Calibration Methods)	1 (Introduction)	17 (Atomic Spectrometric Methods)
Interlaboratory Comparisons	5 (Quality Assurance and Calibrat on Methods)		4 (Good Laboratory Practice: Quality Assurance and Methods of Validation)
Sampling Error and Sampling Design	27 (Sample Preparation)	Appendix 1	26 (Environmental Sampling and Analysis)

Statistics			
Errors in Measurements and their Effect of Data Sets	3 (Experimental Error) 2 (Tools of the Trade)	1 (Introduction)	3 (Statistics and Data Handling in Analytical Chemistry)
The Gaussian Distribution	4 (Statistics)	Appendix 1	3 (Statistics and Data Handling in Analytical Chemistry)
Statistical Tests of Data: The t Test	4 (Statistics)	Appendix 1	3 (Statistics and Data Handling in Analytical Chemistry)
Statistical Tests of Data: The F Test	4 (Statistics)	Appendix 1	3 (Statistics and Data Handling in Analytical Chemistry)
Linear Regression for Calibration of Instruments	4 (Statistics)	1 (Introduction)	3 (Statistics and Data Handling in Analytical Chemistry)
Equilibrium			
The Importance of Ionic Strength	7 (Activity and the Systematic Treatment of Equilibrium)	N/A	6 (General Concepts of Chemical Equilibrium)
The Importance of Ionic Strength – Biochemistry Focus			
Activity and Activity Coefficients	7 (Activity and the Systematic Treatment of Equilibrium)	N/A	6 (General Concepts of Chemical Equilibrium)
Multiple Equilibria	6 (Chemical Equilibrium) 12 (Advanced Topics in Equilibrium)	N/A	11 (Precipitation Reactions and Titrations)
pH of Solutions of Strong Acids and Bases	8 (Monoprotic Acid-Base Equilibria)	N/A	7 (Acid-Base Equilibria)
Acid-Base Distribution Plots	8 (Monoprotic Acid-Base Equilibria) 9 (Polyprotic Acid-Base Equilibria)	N/A	7 (Acid-Base Equilibria)
Acid-Base Distribution Plots-Biochemistry		N/A	
Acid-Base Distribution Functions	9 (Polyprotic Acid-Base Equilibria)	N/A	7 (Acid-Base Equilibria)
The Buffer Zone: What is a Buffer and in what pH range is it effective?	8 (Monoprotic Acid-Base Equilibria)	N/A	7 (Acid-Base Equilibria)
The Buffer Zone – Biochemistry Focus			

When Acids and Bases React: Laboratory	1 (Chemical Measurements) 10 (Acid-Base Titrations)	N/A	8 (Acid-Base Titrations)
Electrochemistry			
Electrochemistry: The Microscopic View of Electrochemistry	13 (Fundamentals of Electrochemistry)	22 (An Introduction to Electroanalytical Chemistry)	12 (Electrochemical Cells and Electrode Potentials)
Electrochemistry: Calculating Cell Potentials	13 (Fundamentals of Electrochemistry)	22 (An Introduction to Electroanalytical Chemistry)	12 (Electrochemical Cells and Electrode Potentials)
Spectrometry			
The Beer-Lambert Law	17 (Fundamentals of Spectrophotometry)	6 (Introduction to Spectrometric Methods) 13 (An Introduction to Ultraviolet-Visible Molecular Absorption Spectrometry)	16 (Spectrochemical Methods)
The Beer-Lambert Law – Biochemistry Focus			
Atomic and Molecular Absorption Processes	17 (Fundamentals of Spectrophotometry)	6 (Introduction to Spectrometric Methods)	16 (Spectrochemical Methods)
Introduction to Spectrometers	17 (Fundamentals of Spectrophotometry) 19 (Spectrophotometers)	7 (Components of Optical Instruments)	16 (Spectrochemical Methods)
Infrared Spectrocopy: Vibrational Modes	19 (Spectrophotometers)	16 (An Introduction to Infrared Spectrometry)	16 (Spectrochemical Methods)
Mass Spectrometry: Interpretation	21 (Mass Spectrometry)	20 (Molecular Mass Spectrometry)	22 (Mass Spectrometry)
Chromatography and Separations			
Introduction to Chromatography	22 (Introduction to Analytical Separations)	26 (Introduction to Chromatographic Separations)	19 (Chromatography: Principles and Theory)
Introduction to Gel Filtration – Biochemistry Focus			

Band Broadening Effects in Chromatography	22 (Introduction to Analytical Separations)	26 (Introduction to Chromatographic Separations)	19 (Chromatography: Principles and Theory)
Gas Chromatography or HPLC? Which do you Choose?	22 (Introduction to Analytical Separations) 23 (Gas Chromatography) 24 (High-Performance Liquid Chromatography)	27(Gas Chromatography) 28 (Liquid Chromatography)	20 (Gas Chromatography) 21 (Liquid Chromatography and Electrophoresis)

D.C. Harris, *Quantitative Chemical Analysis,* 8th Edition, 2009, W.H. Freeman: USA.

Skoog, Douglas A., F. James Holler, Stanley R. Crouch, *Principles of Instrumental Analysis,* 6th Edition, 2007 Cengage: USA.

Christian, Gary D. *Analytical Chemistry,* 9th Edition, 2013, John Wiley and Sons: USA.

Prepared by Elizabeth Jensen, and Caryl Fish

Printed in the USA
K049767SCI032817 01S29053000000001868